Digital Technologies, Temporality, and the Politics of Co-Existence

Mark Coeckelbergh

Digital Technologies, Temporality, and the Politics of Co-Existence

palgrave
macmillan

Mark Coeckelbergh
Department of Philosophy
University of Vienna
Wien, Austria

ISBN 978-3-031-17981-5   ISBN 978-3-031-17982-2   (eBook)
https://doi.org/10.1007/978-3-031-17982-2

Cover pattern © John Rawsterne/ patternhead.com

This Palgrave Macmillan imprint is published by the registered company Springer Nature Switzerland AG.
The registered company address is: Gewerbestrasse 11, 6330 Cham, Switzerland

Contents

1

Introduction: Time, Existence, and Technology

Abstract This chapter introduces the theme of the book and outlines its content. Today we feel like we do not have enough time and that everything is accelerating. We fear death and face an uncertain world. With digital technologies we live too much in the present or perhaps we are not present *enough*: to ourselves, to others, to the world. Digital technologies shape and transform our current vulnerable existence. The book tries to further unpack what this means. For this purpose, it engages with literature that already started to analyze the temporalities created by digital media and responds to thinking about the politics of time. But using process philosophy, narrative theory, and the author's concept of 'technoperformances', it proposes its own original vocabulary to conceptualize how we exist in time and make time today. It also reflects on the normative aspects of digital existence: on the ethics and politics of technoperformances of time. Who and what has the power to shape our time(s)? How can synchronize and co-exist well in the Anthropochrone, and what can we learn from other cultures to answer that question?

Keywords Digital technologies • Time • Temporality • Existence • Vulnerability

We do not have much time. We do not have much time left. We better hurry up. That is at least the message we get from the sciences, from technologies, from media. Our digital clocks and calendars tell us that we need to move on to the next calendar item. The next meeting. The next task. Time is up. Our day is short. We have to work and live fast. We have deadlines. If we do not deal with our emails now, new ones will be added fast and we will lose control. If we do not plan a leisure event or take a course now, we will miss it. If we do not call our friend or our doctor now, it will be too late. If we do not buy this product, we miss out on something. We live in a society of acceleration and speed.[1] Time is scarce, and it becomes even scarcer now that we have digital technologies. As Thomas Eriksen puts it in *Tyranny of the Moment*: we have so much 'time-saving technologies'; and yet 'millions of us have never had so little time to spare as now'.[2] Or as Hannah Arendt already wrote in *The Human Condition*: 'speed has conquered space'.[3]

This feeling of speed and of lacking time also has an existential aspect. Sometimes the seemingly boundless digital time is mercilessly interrupted by the realization of finitude. As we are busy with our phones and computers, the weeks, months, and years pass by, fast and certain. Digital time and space seem to have no limits, but our life time is scarce and we are mortal. Some day our time will be up. Not in general, not in the abstract, but for you, for me. We (will) get older and we will die. That is difficult to accept, especially in a culture that worships youth and encourages us to live as if we will live forever. We fear the future, understood as the end of youth and the beginning of the end. The future of others and our own future. We do not want the future because the future means the end. Once we become aware of our finitude, the future is no longer something abstract or something that humanity is progressing towards; instead, the future gets painfully personal. As the years pass by, we realize

---

[1] A classic work about speed and society is Virilio, *Speed and Politics*. Earlier Georg Simmel analyzed modern temporality in the context of the metropolis, pointing to acceleration and immediacy. For more recent work on speed and acceleration in society see, for example, Rosa and Scheuerman (eds), *High Speed*. Another term that has been used in this context is 'time-space compression', coined by David Harvey in *The Condition of Postmodernity*.

[2] Eriksen, *Tyranny of the Moment*.

[3] Arendt, *The Human Condition*, 250.

that our loved ones will die and that we get closer to our own death. We realize that, as Martin Heidegger observed in *Being and Time*, we experience *Angst*.[4] While in our everyday lives we tend to tranquillize ourselves about it and see death as something that happens in the world—there are cases[5] of death, there are the statistics—suddenly I may realize that the time will come that it is my turn. Dying is then no longer just something "one" does and that happens in the world but something that will happen to *me*—and to take this beyond Heidegger and most other famous existentialists: something that will happen to my parents, my partner, my friends, and so on.

Yet at the same time today, science, technologies, and media do not only distract from awareness of our own finitude but also remind us about it in ways that render it harder to avoid *Angst*, since death becomes more imaginable and concrete. The natural sciences and statistics remind us that our life spans are limited. Data science gives us the numbers. Average life expectancy. Deaths due to pandemic and war. But this is now no longer mere abstract, distant knowledge. Death is shown, for example on TV and social media. This is still about "cases" but the people we see turn out to be *very much like us*. There is still an impersonal aspect. "Humanity" is faced with challenges. We experience the globalization of death, facilitated by high-speed connectivity. The virus flies everywhere. War enters the living room. Speed leads to what Benjamin Bratton calls a 'carnival of accidents.'[6] Paul Virilio famously wrote about the violence of speed.[7] But in addition digital media bring us the violence of violence, and they render that violence more personal. We get the photos in our face. Social media inform us every day about people that pass away in ways that turn abstract cases into personal stories. We see numbers about the COVID-19 pandemic but also the bodies and people who are left behind. We see concrete images of war. We see that people die, and these people are like us or are even directly connected to us. Social media show that people far away die but also friends and colleagues. Social media also

---

[4] Heidegger, *Being and Time*, 235.

[5] Heidegger, 234.

[6] Bratton, "Logistics of Habitable Circulation," 23.

[7] Virilio, *Speed and Politics*, 153.

show us photos of ourselves and others from the past. We see that we were younger and that we get older. These media thus continue to remind us about our personal finitude and mortality.

And amidst all this speed and acceleration, we look for meaning. This is already difficult when we are faced with the scientific information. As Vilém Flusser put it: we are tasked 'to reflect upon the way in which, despite everything, it is possible for human beings to give significance to their lives in face of the chance necessity of death.'[8] How can we combine the temporality of our little human life with the temporalities that science offers: the long temporalities of geological history and climate change, the unimaginable temporalities of the universe? And what about the temporalities of viruses and bacteria? The temporality of weapons and war? The temporalities of financial algorithms? The temporalities of artificial intelligence (AI)? How can we make sense of this? How can we live a good and meaningful life under these circumstances and conditions? How can we live *knowing* all that, or rather, being faced with all these risks, uncertainties, and unknowns? This was already a challenge in earlier modern times. But looking for meaning becomes even more difficult when we are confronted with our finitude in the mirrors and treadmills of social media, which seem to promise the eternal digital life and youth but at the same time remind us that we are done for. The same goes for all the measurements taken from us by health apps and other apps. Death by data. And not to forget: the data and algorithms will be around longer than me. A new type of Promethean shame—to use a term from Günter Anders[9]—is born. I am mortal, but my data will live forever. Not the kind of immortality we had in mind (at least not most of us). Once again, we are humiliated by our own creations. And we are forced to acknowledge our finitude.

The same existential questions are asked with regard to humanity at large, which faces death and extinction. With the help of science and technologies, our days are *counted*. This is partly due to human action. The probability of more and escalating global conflicts is high. Local and regional conflicts may explode into a new world war. The climate is

---

[8] Flusser, *Philosophy of Photography*, 82.

[9] Anders, *Über die Seele*.

changing. Civilizations will disappear. Doom awaits humanity. If climate change does not get the better of us, technologies such as artificial intelligence, bioweapons, or—unfortunately very relevant again—nuclear weapons will wipe out humanity. But even without our doings, the end will come. The sciences tell us that humanity's time on earth is limited. There will be disastrous events, on earth and in the cosmos. Eventually the sun will engulf us and humanity will anyway disappear, unless—so transhumanists argue[10]—we hurry up with technological development, create general AI, upload ourselves to computers,[11] and spread into the universe. What are we waiting for? If we do not move into digital and cosmic space, humanity will disappear like snow in the sun.

While the latter long-term risks this may not lead to *Angst* in a strictly Heideggerian or other philosophical existentialist sense, since that future no longer concerns me (or those that are presently near to me, for that matter), being reminded of all these existential risks—for example via digital social media—leads to a general climate of anxiety and fear. When these other times and time scales come to dominate my life time, my time becomes a pre-disaster time, a time before the end. As Extinction Rebellion's white-faced, veiled figures dressed in scarlet red robes perform a funeral procession for the planet and its beings in order to warn about climate change,[12] we realize that this is not only about "the earth" but that it is also our blood that is mourned. The planet's widows and widowers (or ghosts?) seem to visit us from the future to tell us that the end is near, including the end of humanity. Digital technologies and media seem to at once deny all this by promising us a transhumanist, post-earth future *and* show us forcefully and undeniably that our future is in danger. It is almost too late. Things are accelerating. But we can still do something, at least if we act fast. In that sense, both the current system and its protesters are part of the same global empire of speed, the same pervasive techno-chronic regime.

---

[10] See for example Tegmark, *Life 3.0*.

[11] Kurzweil, *The Singularity is Near*.

[12] See for example this photo of Extinction Rebellion activists in London in 2019. https://www.bbc.com/news/uk-49957521

However, most theories of time and digital technologies tell us that digital technologies do not only accelerate our lives but also tie us to the present. They warn of *presentism* or what Paul Virilio calls 'presentification'[13]: we focus on the immediate and thereby risk to lose the past and the future. Manuel Castells uses the term 'timeless time': in the network society, we find ourselves 'in perennial simultaneity'.[14] Time is compressed and flat. One could say that there is no narrative time; there is no (hi)story, only immediacy. Bernard Stiegler, who recognized that human time is technologically shaped, also argued that technologies are designed to focus our attention on the present.[15]

This has implications for how we experience existence and live our lives. Heidegger argued that we live inauthentically if and when we do not confront our being-towards-death.[16] By relating to the future possibility of death, we can overcome being in the present. According to him, being in the present is inauthentic. Digital technologies seem to contribute to this inauthenticity in so far as they focus on the present and make us forget about the future, in particular the possibility of our own death. One could add that presentism also makes us forget about the future of the climate and the earth. Presentism literally has no time for longer histories, be it human histories, climate histories, or geological histories. And if there is time for the climate at all, it is time for the "climate present" in the form of a crisis, a protest, a conference. The temporal horizon of the longer, geological and climate timescales of the past and the future is not directly reachable and can only appear in the present as crisis, countdown clock, and deadline within the frenzy and anxiety of a hectic digital now.

That now becomes a distraction. What Walter Benjamin notes about film is also applicable to social media: we expose ourselves to 'shock effects' in order to cope with the dangers that threaten us.[17] We let ourselves be bombarded by spectacular news on digital social media. We try

---

[13] Virilio, *Politics of the Very Worst*, 81. See also Guerlac, *Thinking in Time*.

[14] Castells, *Network Society*, xli. For an overview of some work on digital media and the present, see for example Coleman, "Experiencing 'the Now'."

[15] Stiegler, *Technics and Time III*. For a critical discussion, see Paris, "Time Constructs."

[16] Heidegger, 234.

[17] Benjamin, *Illuminations*, 229.

to cope with the risks by taking distance and distract ourselves. Digital media, known for destroying our focus and attention,[18] help us with this. As the 21st equivalents of Benjamin's *flaneurs*,[19] we try to take distance from the world around us by immersing us in the nonchalance and indifference to time that digital media and virtual environments seem to offer. Digital media and technologies enable us to escape to a seemingly timeless world or to a shocking, brutal now in order not to think about the dangerous and deadly futures that await us. But this brings only temporary relief. Death is certain and total destruction a real possibility. Moreover, globalization, combined with increased speed, also means increased vulnerability. As we are hyper-connected, we are also dependent and vulnerable in relation to distant people and events. The Coronavirus pandemic shows how globalized we already are. Vulnerability is a matter of time and speed. The time it takes to get infected. The speed of vaccination development. The timeliness of the measures taken. Our bodies become sites where global pandemic forces, struggles, and indeed *races* play out, for example, between a particular virus variant and the pharmaceutic industry, between the behavior of the population and the government. There are also races about developing the best and fastest advanced technologies, attaining the most geopolitical power, winning terrain in a war, and so on. Digital technologies support these races, and the races are often about digital technologies, for example AI. Countries race to have the best AI, the fastest AI. These races about speed and acceleration are further supported by capitalism and neoliberalism. While climate change is already undermining the prevailing system, speed capitalism still reigns, on the financial markets (consider high frequency trading), in technology development, and elsewhere. Robert Hassan speaks of socially and technologically created 'empires of speed': technologically based forms of time, including computer-based 'real-time', dominate other forms of time.[20]

---

[18] For some discussion, see for example Lee, "Coexistence."

[19] Judy Wajcman and Nigel Dodd point to the interesting relation to time and the world that the flaneur has: he 'removes himself from the fast-moving world while simultaneously immersing himself within it'. Wajcman and Dodd, *The Sociology of Speed*, 18.

[20] Hassan, *Empires of Speed*, 3.

There is some truth in these claims about immediacy. It seems that digital technologies make us focus on real-time, on the now(s) of digital interaction. Despite latency[21] and lags[22] that occur between our embodied interaction and their desired effect, we are interacting with digital technologies as if our lives depend on it: the interaction is happening in an intense, hungry, and expansive digital now—regardless of what our bodies and minds happen to want and regardless of where our bodies are situated in space. And perhaps today our lives *do* depend on it. With digital time and digital technologies we work hard to fight against our vulnerability,[23] and when we lose ourselves in social media, we might temporarily forget about our own finitude. We also forget about the future of the earth. As we are processed and distracted by digital technologies, we do not see the processes beyond the intensive presents in which we immerse ourselves or beyond the stressful (self-)presentations we perform. We live and post against the clock. We also work against the clock. In capitalist contexts, immediate profit prevails over the future of pensions, the future of living beings, and the future of the planet. Digital technologies and media seem to support this. The data factories work 24/7, and all of us are its laborers. The digital present rules. Seen from this angle, the present and its digital technological allies are highly problematic. This is not the kind of present most of us (the 99%?) want.

But the opposite can also be argued: influenced by Buddhism and mindfulness, one could argue that there is also a sense in which the present and being in the present is good. One could argue that we should be present and attend to the present moment without obsessing about the past and future, but that digital technologies take us out of the present, keep us from being present to ourselves and others, and make us victims and prisoners of the technologically shaped past and future. In our over-organized digital lives, we are always looking ahead to what comes next, and we are pinned to our past in various ways, for example to our social media profile or to the profile AI made of us based on past data. As AI is trained on data from the past, that past shapes the present and the future.

---

[21] Floridi, "Digital Time."

[22] Sebald, "Loading."

[23] Coeckelbergh, *Human Being @ Risk*.

Do we still have a future? We are oriented towards the future, but our future is technologically fixed and predicted. And this is not just done by the algorithms, humans also have a role in it: we start to create that future as we behave according to what is technologically predicted. As Helga Nowotny has warned, we begin to live the future of the predictions made by algorithms. If we believe that it is the only possible future, 'we risk closing down other options'.[24] The present is hijacked by the future, and it is not an open one. In these ways, we are not really in the present: we are not mindful,[25] we are restless and distracted, and we live the closed future made by AI. In this sense, both past and future become a prison, and a return to the present is then a liberation.

Living in this intense compressed time, in this mediated present, or indeed in the *non-present* of digital technologies, is tiring and emotionally burdensome. The constant anxiety and temporal captivity are stressful, also existentially speaking. Especially since as users of digital technologies we are all the time reminded about our finitude and impermanence. On social media we are anxious about our finitude, not only because we are notified that others die and see how we get older and hence move closer to our own death, but also since *we constantly die digitally*. We literally click and scroll ourselves to(wards) death. What is on the screen at one moment is no longer there the next moment. Images appear and go again, including images of ourselves. Digital existence, like all existence, is unstable and vulnerable. We are made into data and data flow. Data come and go. Whereas industrial technologies with their artefactual, thing-like character could still give us the illusion of permanence, the digital shows us unavoidably that nothing is permanent. Data are immaterial. They exist but they are no-thing. As data resources and data points, we too are no-thing. It seems that we are as impermanent and mortal as the brief flickering of a number on the screen, the sudden and transient appearance of an image on the timeline of the world. This is difficult to accept, especially for the Western mind.[26]

---

[24] Nowotny, *In AI We Trust*, 35.

[25] David Kreps and colleagues argue for what they call 'IT mindfulness'. Kreps, Rowe, and Muirhead, "Understanding," 6140.

[26] Buddhists would have less problems with impermanence and no-self.

We struggle against death and impermanence. And digital social media even asks us to, in the sense that we continuously have to be present in order to continue our digital lives, in order to not die digitally. We have to continuously present and perform ourselves online while not being really present to the situation. We have to perform digitally or die. Post or perish. We have to be and stay online. We have to interact. We have to produce data. We even take selfies in order to try to maintain ourselves and our presence. But these efforts are in vain. We are not in the present and we cannot hold on to the present. There is no time for the present. We and our time are resources for the future. The future is being calculated. The future is made by algorithms and those who use them to use us. And with AI and data science, we are tied to our past data and fixed to anticipatory logics. Even ethics and regulation of technologies tend to often miss the situatedness, the contexts, and societal circumstances of the making of data and 'the contingencies of living particular lives'.[27] AI ethics becomes itself an algorithm. Everything has already been predicted. We want a future, but the future is closed. Our data have already been taken. We have been analyzed, modelled, and profiled. Our future is fabricated. Our lives are predicted. Uncertainty has been managed. And probably we are being manipulated right now. We do what the algorithms predicted. We are not really in the present but are fixed in the past and closed off to the future. We are not authentic at all since we miss the present and are lost in the flow of presentation. It is impossible to reflect on the possibility of death if you are continuously dying, if you are continuously datafied, made into a thing.

These claims about digital technologies and time are possible and make sense at all because digital technologies, like all technologies, are not mere instruments but have unintended effects: effects on what we do, on what we are, and on what we want to become. They shape our experience and perception, our knowledge, our relations to others, and our society and culture as a whole. As Marshall McLuhan famously said: the medium is the message.[28] For example, digital technologies and media have the effect that they further globalize us and our vulnerability. The globe becomes

---

[27] Pink and Lanzeni, "Future Anthropology," 3.
[28] MacLuhan, *Understanding Media*.

our skin as we are extended via the Internet and its media. Technologies and media have existential effects. They shape our lifeworld and how we are in the world. They should not be seen as separate domains. Our lives are shaped by them. They are part of our world and we are part of their world. As Amanda Lagerkvist puts it: 'our screens and devices have become part of our existence, with all its inevitable quests for meaning and value'.[29] This means that we can ask existential questions in relation to the digital. The being-in-the world and the thrownness Heidegger writes about in *Being and Time* is now also digital.[30] What it means to be human and mortal today, Lagerkvist writes in *Existential Media*, 'are deeply enmeshed in our technological culture'.[31] For example, coping with loss of a loved one,[32] struggling with disease and pandemic, or experiencing the limits of the earth's ecosystem are now mediated by digital technologies. However, the existential dimension of technologies and media is not only about death, crises, or transformative moments—situations in which, in sharp contrast to the endlessness and 'no limits' the digital seems to promise, we experience limits, finitude. Our lives and existence in general have become at least partly digital. Digital media have become what John Durham Peters calls the 'infrastructures of being, the habitats and materials through which we act and are'.[33] Or put in the language of vulnerability and risk I proposed in *Human Being @ Risk*: engaging with digital technologies is now a way of being-at-risk, a way of 'being-vulnerable'.[34] Our existential condition as a whole, which is always one of being vulnerable and being uncertain, has been significantly transformed by digital technologies. And, important for our current topic: digital technologies shape temporality. They shape our relation to time. Stronger: they shape our time and life time.

---

[29] Lagerkvist, "Digital Existence," 1.

[30] Lagerkvist, "Existential Media."

[31] Lagerkvist, *Existential Media*, 1.

[32] Lagerkvist discusses such 'digital limit situations' in *Existential Media*. For example, in Chapter 5 she describes how grieving people relate to the ethos of quantification as they look for consolation in likes on their digital commemoration site.

[33] Peters, *The Marvelous Clouds*, 15.

[34] Coeckelbergh, *Human Being @ Risk*.

It was already known that clocks and computers are not just neutral instruments that help us to tell time and calculate, but also shape human experience and culture. If we have the feeling of speed and acceleration, this is in part due to the technologies we use. As David Bolter put it in *Turing's Man*: 'There is an intimate connection between a culture's attitude toward time and the technology by which it measures time'.[35] Bolter and others such as Lewis Mumford[36] showed that computers, and earlier the mechanical clocks and the mathematization of time, already influenced how we think of time and progress and have a complex relation to human experience and human lifeworlds. Time became more abstract, and—at least according to Bolter and many philosophers in the phenomenological tradition from Husserl and Heidegger until today—time also became removed from the lifeworld and from the cycles of biology and nature. Life became faster, too fast. Time became a commodity. The idea of unlimited progress emerged, as did the idea of perfecting human beings.[37] The challenge then is to re-connect time to life and the lifeworld.

However, going beyond this classic phenomenological philosophy of technology tradition that conceptually keeps apart lifeworld and technology, and instead keeping more in line with contemporary philosophy of technology such as postphenomenology, one could also argue that time remains part of the lifeworld, but that both lifeworld and time are being transformed by technologies and media in ways that shape our experience in particular ways—*one* of which may be that we experience being alienated from the cycles of nature, but there are also other possibilities. Moreover, instead of speaking of *one* time it seems more appropriate to speak of different times and temporalities. For example, there is clock time but also the time of the sun and the time of the seasons. There are also cultural differences: as I will stress later in this book, different cultures mean different temporalities: different ways of relating to time.

Today, all kinds of digital technologies shape our temporalities and, in the end, our existence. For example, Judy Wajcman has argued that

---

[35] Bolter, *Turing's Man*.
[36] Mumford argued that in the monasteries of the West, clocks were invented to exercise order and discipline. Mumford, *Technics and Civilization*, 13.
[37] Bolter, *Turing's Man*, 123.

digital calendars, tied to the Silicon Valley ideology of optimization, have further encouraged clock time.[38] And recently I have argued that AI, mediated by narratives, shapes our time and lifeworld by connecting our past, present, and future in specific ways.[39] In interaction with a particular societal and cultural contexts,[40] digital technologies influence how we live and how we think about climate change and about the future of humanity. More generally, AI and digital technologies shape how we exist today and how we *time* and *are timed* today.

This book tries to further unpack and conceptualize what this means: what it means to say that digital technologies shape our relation to time and our existence, and what this implies for existing in times of climate change and new technological possibilities such as AI. It offers some concepts that help us understand how digital technologies and media enable and support particular ways of relating to time and existence. Engaging with literature that already started to analyze the temporalities that are created by digital media, it creates its own conceptual language to articulate this usually unintended influence. In particular, it is argued that we can understand this influence in terms of process, narrative, and performance. In response to classic phenomenology of time and technology, it proposes a less dualist and more relational approach to time, technology, and human being, which does not oppose time technologies to the lifeworld, as Bolter and so many others did, but tries to understand how these technologies shape the lifeworld and, ultimately, also our selves and our existence. For this purpose, the book brings together the following conceptual building blocks. First, using *process* philosophy, it shows a way out of what is today seen as overly deterministic thinking about technology and time,[41] and clears the road for thinking about digital technologies and digital selves not as objects but as processes and becoming.

---

[38] Wajcman, "Digital Architecture."

[39] Coeckelbergh, "Time Machines."

[40] Mumford argued that technologies can only exert their influence in a particular cultural context. He writes: 'The world of technics is not isolated and self-contained: it reacts to forces and impulses that come from apparently remote parts of the environment'. Mumford, *Technics and Civilization*, 6. In this book I radicalize this claim by using process philosophy, narrative theory, and the concept of performance.

[41] See for example Judy Wajcman's criticism of the technodeterministic tendencies in Castell, Virilio, and other philosophers of acceleration. Wajcman, "Digital Architecture," 317.

Second, using *narrative* theory, it helps to conceptualize the structure of digital existence in a way that places technology not only in a temporal but also in a social and cultural context. And third, using the concept of *performance*, it emphasizes the role of humans and the social but also bodily and kinetic character of our (co-)existence with digital technologies. These building blocks suggest an integrated framework that can help us to understand how we exist and relate to time today with digital technologies and media.

Yet the book goes further than trying to understand. It also questions and asks what *should* be done. It asks what philosophers call *normative* questions. Based on process thinking, it asks questions about responsibility and existence. It points to the ethical but also the social and political dimensions of digital technologies. It asks how we *should* shape our technologically mediated existence and, in particular, *how we should relate to time*. Are the current ways we relate to time through digital technologies good ways? Responding to the ancient question regarding the good life, it asks the question about *good times*. What are good times? How can we find ethically good ways of relating to time, given digital technologies? Should digital media re-present, in the sense of bringing back the past, or is that not the kind of present we need? What *is* a good way of relating to time? Should we slow down? Should we find a different relation to time altogether? And what does this mean at the collective level? The book ends with a reflection on the politics and power of relating to time through digital technologies. It asks how we can find ways of good co-existence, at the local and at the global level, and what the political challenges are when we attempt this. Who and what currently has the power to shape our (relation to) time? Who (co-)shapes our processes, narratives, and performances with technology? Who *should* have that power? Can we *synchronize* in different ways than those prescribed by technocapitalism and technocolonialism? Can we find, create, imagine, and have a good *common* time, given digital technologies but also in the light of current global challenges such as climate change and the pandemic? What does it mean to co-exist (well) in what I will call the "Anthropochrone"? And what can we learn from other cultures with regard to common time and coexistence?

References

Anders, Günther. *Die Antiquiertheit des Menschen*. Vol. 1, Über die Seele im Zeitalter der Zweiten Industriellen Revolution. Munich: CH Beck, 1956.

Arendt, Hannah. *The Human Condition*. Chicago: University of Chicago Press, 1958.

Benjamin, Walter. *Illuminations*. Translated by Harry Zohn. Boston: Mariner Books, 2019.

Bolter, J. David. *Turing's Man*. Chapel Hill, NC: The University of North Caroline Press, 1984.

Bratton, Benjamin H. "Logistics of Habitable Circulation: A Brief Introduction to the 2006 Edition of *Speed and Politics*." In *Speed and Politics*, Paul Virilio.

Castells, Manuel. *The Rise of the Network Society*. 2nd ed. Malden, MA: Wiley-Blackwell, 2010.

Coeckelbergh, Mark. *Human Being @ Risk*. Dordrecht: Springer, 2013.

Coeckelbergh, Mark. "Time Machines: Artificial Intelligence, Process, and Narrative." *Philosophy and Technology* (2021). https://link.springer.com/article/10.1007/s13347-021-00479-y

Coleman, Rebecca. "Making, Managing, and Experiencing 'the Now': Digital Media and the Compression and Pacing of 'Real-Time'." *Communication & Sport* 22, no. 9 (2020): 269-298.

Eriksen, Thomas Hylland. *Tyranny of the Moment*. London: Pluto Press, 2001.

Floridi, Luciano. "Digital Time." *Philosophy & Technology* 34 (2021): 407-412.

Flusser, Vilém. *Towards a Philosophy of Photography*. London: Reaktion Books, 2000.

Guerlac, Suzanne. *Thinking in Time*. Ithaca, NY: Cornell University Press, 2006.

Hassan, Robert. *Empires of Speed*. Leiden: Brill, 2009.

Heidegger, Martin. *Being and Time*. Translated by Joan Stambaugh. Albany, NY: State University of New York Press, 1996.

Kreps, David, Franz Rowe, and Jessica Muirhead. "Understanding Digital Events: Process Philosophy and Causal Autonomy." In *Proceedings of the 53rd Hawaii International Conference on System Sciences*, 2020.

Kurzweil, Ray. *The Singularity is Near*. New York: Penguin, 2005.

Lagerkvist, Amanda. "Existential Media: Toward a Theorization of Digital Thrownness." *New Media & Society* 19(1) (2017): 96-110.

Lagerkvist, Amanda. "Digital Existence." In *Digital Existence*, edited by Amanda Lagerkvist New York: Routledge, 2019.

Lagerkvist, Amanda. *Existential Media*. Oxford: Oxford University Press, 2022.

Lee, Sunji. "Coexistence between attention and distraction." *Educational Philosophy and Theory* 54, no. 5 (2022): 512-520.

MacLuhan, Marshall. *Understanding Media*. New York: Mentor, 1964.

Mumford, Lewis. *Technics and Civilization*. New York: Harcourt, Brace and Company, 1934.

Nowotny, Helga. *In AI We Trust: Power, Illusion, and Control of Predictive Algorithms*. Cambridge: Polity Press, 2021.

Paris, Britt S. "Time Constructs." *Time & Society* 30, no. 1 (2021): 126-149.

Peters, John Durham. *The Marvelous Clouds: Toward a Philosophy of Elemental Media*. Chicago: The University of Chicago Press, 2015.

Pink, Sarah, and Debora Lanzeni. "Future Anthropology Ethics and Datafication." *Social Media & Society* (2018): 1-9.

Rosa, Hartmut, and Scheuerman, William E., eds. *High Speed Society*. University Park, PA: The Pennsylvania State University Press, 2009.

Sebald, Gerd. "'Loading, please wait' – Temporality and (bodily) presence in mobile digital communication." *Time & Society* 29, no. 4 (2020): 990-1008.

Stiegler, Bernard. *Technics and Time III*. Palo Alto, CA: Stanford University Press, 2010.

Tegmark, Max. *Life 3.0: Being Human in the Age of Artificial Intelligence*. London: Penguin Books, 2018.

Virilio, Paul. *Politics of the Very Worst*. Translated by Michael Cavaliere. New York: Semiotext(e), 1999.

Virilio, Paul. *Speed and Politics*. Translated by Mark Polizzotti. Los Angeles, CA: Semiotext(e), 2006.

Wajcman, Judy. "The Digital Architecture of Time Management." *Science, Technology, & Human Values* 44, no. 2 (2019): 315-337.

Wajcman, Judy, and Nigel Dodd, eds. *The Sociology of Speed*. Oxford: Oxford University Press, 2016.

2

Process, Narrative, and Performance: Conceptualizing How Digital Technologies Shape Temporality and Existence as Technoperformances of Time

Abstract We tend to see the world as a collection of things, but drawing on process philosophy (especially Bergson), this chapter proposes that we instead think about the world and the self in terms of processes and becoming. The influence of digital technologies on what we are and what we become is then also a matter of process, from which both subjects and objects emerge. We participate in this ontogenesis. Yet the good news is that we are not alone; our digital existence and self are social and relational, for example they have a narrative structure. Ricoeur argued that narration helps us to make sense as it draws characters and events into a meaningful whole. Digital technologies assist and shape this process of narration and meaning-making as co-authors of our narratives. Moreover, it is shown that the concept of performance helps to further develop this process-oriented and narrative view of how technologies *and* humans shape our time and existence: we are not just passive, but also perform these processes and narratives and perform time. The chapter proposes the concept of technoperformances of time to show how humans play an active, social, and moving-bodily/embodied role as they move through time and perform time, without full control but remaining responsible for the meanings created. Digital technologies play an important role in these performances of time and this making of meaning. This also has a political dimension: through our performances we exercise power. Furthermore, there are multiple temporalities.

Keywords Process philosophy • Bergson • Narratives • Meaning • Ricoeur • Performance • Technoperformances of time

We tend to think about the world as a collection of things. We see technologies as things: things that we made ourselves, artefacts (human-made things). Data are also pictured as things. We tend to see the self as a kind of thing: a stable, well-defined thing called 'I'. And even time itself is seen as a succession of present moments, which are imagined as a kind of things (dots, points) situated on a time line. When we try to conceptualize how digital technologies and media shape temporality and existence, then, we tend to imagine how things shape things: how technologies as objects relate to subjective time, and how it shapes that thing called "self" or that "subject"-thing. For example, if we claim that media hurry us or that science reminds us that our life time is limited, then we imagine that they present us with a time line, including—sometimes literally—a deadline. This objective time is then thought to give rise to subjective experience, feelings, and imagination. We imagine that the thing called 'mind' is influenced by objective time. We imagine that the thing called 'I' or 'self' is going to be annihilated. We fear our death and that of our civilization. We imagine that technologies shape us like a hammer hits a nail. Object to object, thing to thing. A kind of Humean billiard-ball picture.[1] Digital technologies make us into a collection of data. I am a collection of things. But, so it seems, before this influence of technologies I was already a thing, just another thing: a fixed thing, an essence, a self, a stable, transparent, and well-defined me. Even events related to climate change are imagined as things—things that threaten us, for example a flood or a storm.

---

[1] Already during and after the second world war, Elisabeth Anscombe and other female philosophers reacted against such a picture of the universe, see Lipscomb, *The Women*, and in particular Anscombe, "Modern Moral Philosophy."

Contemporary philosophy of technology mirrors this common way of thinking about ourselves and technologies to the extent that it focuses on what artefacts, *things*, do.[2] For example, in the postphenomenology of Don Ihde and Peter-Paul Verbeek, things are seen as standing in between the 'I' and the world. Graham Harman's object-oriented ontology,[3] which tries to conceptualize the autonomy of objects and is sometimes also applied to technology, is a metaphysics of objects, things. And the language of 'assemblage', which was introduced by Gilles Deleuze and Félix Guattri[4] and is popular in science and technology studies and related fields,[5] makes us imagine technologies as collections and multiplicities of *things* and humans (who are in turn understood as *parts* of the assemblage). It seems that much contemporary thinking about technologies, media, and humans is all about what things do to other things.

Yet there are ways of thinking about relating to the world, to technologies, and to ourselves that offer a different perspective on what is happening and on what we experience, including a different perspective on meaning and existence. I propose to describe this by using the concepts of process, narrative, and performance.

Process

In the philosophical tradition we find the view that the world is not a collection of things or substances but a *process* or processes. Consequently, what we 'are' is a matter of what we become. Technologies, we could add, shape this process and are part of this process. Moreover, technologies are processes themselves. They also become.

Let me unpack this.

As a view of the self, we find process thinking to some extent in existentialism, for example in the work of Jean-Paul Sartre, who argued that

---

[2] The empirical turn in philosophy of technology meant a turn towards things. See for example Verbeek, *What Things Do*.

[3] Harman, *Object-Oriented Ontology*.

[4] Deleuze and Guattari, *Mille Plateaux*.

[5] For example, Donald MacKenzie uses the term in his material sociology of markets. See for example his "Capital's Geodesic."

in contrast to things such as a paper knife or a cauliflower we are not an essence; 'existence precedes essence'.[6] The self is then a project, which can interpreted as a kind of becoming. There are also some elements of process thinking in German idealism, pragmatism, and phenomenology. But the most radical formulation, at least in Western modern philosophy, is *process philosophy*. According to process philosophy, 'everything' (when we try to formulate this, it is difficult to escape thing-thinking: we are used to think about the world in terms of *things*, thing-thinking is part of our everyday language) is a matter of process: not only are humans not things, but *things are also not things*. Following in the footsteps of the ancient Greek philosopher Heraclitus, who is known for the thought that everything flows (*panta rhei*), process philosophers Henri Bergson[7] and Alfred North Whitehead[8] propose to see everything, including human existence but also that what we call 'things', in terms of processes of becoming. Whitehead says that 'the actual world is a process' and the process is the 'becoming of actual entities' or 'occasions'.[9] Reality is a matter of becoming, not of things. Humans and their selves also become. We are made in processes, and we make things in processes. The self flows. It is like Heraclitus's river.

Process thinkers criticize the split between objective and subjective that tends to be assumed and performed in modern thinking. Subjective experience is not something separate or virtual but is also real and makes things what they are. Both subject and object are continuously created and continuously co-emerge in processes. We can apply this view to our relation to technologies and media: technologies and human experience co-emerge in the process of use. We can also apply it to science and epistemology. Modern thinking opposed the reality of the world to the virtuality of consciousness. But according to process thinkers, we cannot even know the real, understood as what Bergson calls duration, without involving subjective experience. Bergson writes that 'we cannot speak of

---

[6] Sartre, *Existentialism is a Humanism*, 22.
[7] Bergson, *Creative Evolution*.
[8] Whitehead, *Process and Reality*.
[9] Whitehead, *Process and Reality*, 22.

a reality that endures without inserting consciousness into it'.[10] Once we speak of a before and after, we have already inserted consciousness. The real includes consciousness. As David Kreps and colleagues summarize process philosophy's world view: 'the universe is making itself up as it goes along—through our consciousness'.[11] One can also use again the term 'becoming' to repair the subjective/objective split: both we and things become. Similarly, process philosophers question the distinction between objective and subjective time. We are made in time and by time; we become. Therefore, it does not make sense to distinguish between objective and subjective time, time as such versus time for us. Instead, time needs to be understood non-dualistically. Time is duration and becoming.

Process philosophy has not had much influence on philosophy of technology and media, except in France. Gilbert Simondon and Bernard Stiegler, and to some extent Gilles Deleuze and Bruno Latour are the key figures here: each in their own way was interested in processes, in how things come into being, rather than on what something is. For example, Simondon used the concept of 'individuation' to conceptualize an ontogenesis (rather than an ontology) of psychic and technical elements. In *On the Mode of Existence of Technical Objects,* he wrote about the genesis and concretization of technical objects in order to conceptualize 'the integration of the reality of technics into culture'.[12] Technology, understood in terms of process, evolution, genesis, and concretization, also shapes culture. Once we look at what happens on the level of human practice, subject/object dualism is dissolved; both human beings and technical objects are expressions of life.[13] In other words, Simondon's philosophy of technology is a philosophy of becoming, which includes a non-dualist view of our relation to technology. Later Stiegler, influenced by Heidegger but also by Simondon, was sympathetic to this ontogenetic, process approach to technology and proposed to create new political processes, although it is questionable if he really managed to overcome the subject/

---

[10] Bergson, *Duration and Simultaneity*, 48.
[11] Kreps, Rowe and Muirhead, "Understanding Digital Events," 6141.
[12] Simondon, *Technical Objects*, 176.
[13] Schick, "The Potency of Open Objects."

object split. He assumed the co-constitution of human interiority and technical exteriority and the co-evolution of humans and technology, but did not go all the way to embrace process philosophy's non-dualist outlook.[14]

For thinking about digital technologies and existence, process thinking implies that what digital technologies do to us is not a matter of objects doing things to subjects. Rather, both subjects and objects are made and emerge in digital processes and digital events, which are also processes of generating meaning (i.e., the making and becoming of meaning) and which are of existential significance. According to this interesting non-dualist view, technological-existential processes and events are both nature and culture, object and subject, personal and more-than-personal, scientific and existential, consciousness and non-consciousness. Kreps defines digital events as 'concrete slabs of existence defined by a term or period, in which all physio-chemical processes and personal subjective experience are included'.[15] One could say that both the technologies and its human users *become* in these digital processes and events. There are not first subject and objects, which then somehow interact and influence—perhaps even co-constitute[16]—each other. Engaging and interacting with digital technologies is an existential and dynamic process in which what the human users become and the technologies become *emerge* from that process. In digital technological processes both subjects and objects are created and emerge. They are generated in processes of interaction, in encounters.

Furthermore, human existence itself is a process of becoming and emergence, in which there is not an isolated, alienated consciousness standing apart from the world which is then (often rather harshly) confronted with the world, as in Western existentialism, but in which existence, meaning, and world emerge and are constituted in and by a process. Technologies participate in this process of meaning making, emergence, and becoming. Process thinking also delivers a relational view of

---

[14] For more discussion see Rambo, "Technics Before Time."

[15] Kreps, *Understanding Digital Events*, 5.

[16] Verbeek argues that when technologies media our relation to the world, we are co-constituted by this mediation. But he does not go all the way to a process view of what happens.

existence with technologies, since what we become is always a becoming-in-relation: we become in interaction and interrelation to others and to technologies.

When we want to know about how digital technologies and media impact temporality, then, the question to ask is not what the technologies do to objective time as opposed to subjective time, but rather to figure out how the technologies shape and participate in these relational, existential, and emergent processes and events and how in these processes both time and technologies are *themselves* the emergent result of processes, events, and interactions. How we experience time when we interact with digital technologies is not to be distinguished from objective time, but is a lived time and time as duration, a temporal process that encompasses both the subjective and the objective, the cultural and the technological, the human and the non-human. In this process, consciousness is mixed with, and interacts with, the material and technological realities. Digital experience and digital processes are entangled. My self and existence are made, lived, and emerging from that experience, interaction, process, and becoming.

For example, when we feel hurried as we use digital technologies and anxious in the light of digitally mediated events and processes such as climate change, pandemic, and war, then according to process philosophy this hurriedness and anxiety is not just a subjective feeling that can be 'corrected' or 'verified' by referring to objective time. Instead, if we apply the concept of what Bergson calls *duration*, then what is happening here really amounts to a change of time and a change of both subjects and objects. Subjects and objects co-constitute one another. Digital technologies become machines of speed, and we become hurried and anxious selves. Climate change, perceived via digital media, is an event or a set of events, which include subjective and objective aspects, human and non-human aspects, and digital and analogue aspects.[17] In the process, experience and technologies emerge and merge, and a particular form of human existence takes shape. In the context of climate change, technologies of

---

[17] As Walter Zimmerli has rightly argued, the analogue is digitalized but the digital also has to be transformed back into analogue elements. See Zimmerli, "Künstliche Intelligenz," 18. One could say that there is a continuous process of digital-analogue and analogue-digital transformations.

measurement become technologies of doom, and humans are made into subjects-in-crisis. The same happens in and by pandemics and wars, which are also shaped by digital technologies and by data, and which turn digital technologies and media into fear machines and create pandemic subjects and war subjects. The point is not that a particular digital technology is a medium, in the sense of a separate thing 'in between' us and reality, which then shapes our time and consciousness. Rather, human subjects, reality, and indeed our digital media become what they become *in the process.*

It is clear that in a weak sense digital technologies can be understood as processes: they are technological processes. For example, data are made in, and emerge from, data science processes. AI can be understood as a process including data collection, creation of algorithms and models, data analysis, and so on. In these processes, humans play a key role, for example as designers and developers but also as those who maintain the technologies and infrastructures. Not only cars but also digital technologies need maintenance: software needs updates and cloud infrastructures need maintenance.[18] Technologies are not only created, designed, and developed; they also have life spans or life cycles, and at different moments in their 'life', technologies are connected to different practices. Looking at digital technologies in terms of process already opens up an interesting *temporal* perspective on technologies that goes beyond seeing them as things, as artefacts at one moment in time, and enables us to ask for example questions regarding the sustainability of digital technologies.

But drawing on process philosophy also enables us to tap into a deeper and stronger meaning of process, according to which humans and technologies, subjects and objects, change and mutually constitute one another. What data and the related digital technologies are and indeed what *we* are *changes* in the process. And thus the outcome of the process also changes and emerges. In times of crisis, digital media co-create the crisis and at the same time, that is, in the process, these media become anxiety machines. In the same process, we—interacting with these machines—become anxiety subjects. There are no fixed essences but

---

[18] Maintenance is an often neglected topic in contemporary philosophy of technology. Mark Thomas Young and I are currently trying to change that.

objects and subjects in inter-relation, flux, and evolution. Our relation to time, our temporality, is also changing and gets shaped in the process. For example, in the light of climate change as mediated by science and digital technologies, my life time becomes part of a climate history, present, and future. My existence becomes 'climatized', so to speak. I am now a different subject than before. The same happens in pandemic times, times of war, etc.: my existence and subjectivity becomes pandemic, war, and so on. How I experience and live my time becomes shaped by these processes and the related temporalities, for example geological time, war time, or the temporality of the spread of a virus. In these times, we become climate subjects, pandemic subjects, and war subjects. Our existence changes. But in the process, the objects also change. Digital technologies are part of these processes and events; they contribute to the process and are in turn shaped by them. Digital media become climate media, pandemic media, war media. The Internet today, after the pandemic, is no longer the same Internet as before the pandemic.

A process philosophy perspective thus also offers an interesting view of technological mediation that renders it temporal, that moves it. Instead of seeing mediation as a *milieu* in which the subject finds itself and that is external to us, or as something that sits 'in between' subjects and their world, as postphenomenologists Don Ihde[19] and Peter-Paul Verbeek[20] present it in their work, mediation is understood as a process. Timothy Barker puts it as follows: 'Mediation is not a flow between two preexistent entities; rather, it is a process that re-presents or reconstitutes entities. In short, it is a generative process, setting the conditions for the becoming of entities. This is a temporal process, with technological processes generating particular conditions for becoming'.[21]

Digital technologies as they are used and understood today, however, risk to present a picture of the world that is made of things and an image of a self and existence that is fixed, that is not really a *becoming*. They also deny the deep relationality of both world and self. As we are presented as things and made into things (and indeed make ourselves into things

---

[19] Ihde, *Technology and the Lifeworld*.
[20] Verbeek, *What Things Do*.
[21] Barker, *Time and the Digital*, 12.

through the technologies), we forget that we change in relation to the world and in relation to others. Kreps has argued that social media profiles 'give an abstractly concrete and stable identity to identities that are in processes of becoming', that with the digital traces we leave in digital environments we lose the right to be forgotten, and that we are fixed in data systems used for marketing in the context of what Shoshana Zuboff calls surveillance capitalism.[22] Like with AI, social media thus bind us to the past and present a future that seems to be fixed, thereby denying becoming. Moreover, in its discourse on the technological future, transhumanism presents an ontology and a philosophical anthropology that is 'closed'. The world is predictable; human beings are imperfect machines that need an upgrade and AI determines the foreseeable future. We are on our way to the Singularity; resistance is futile.[23] This view contrasts with what Johannes Schick, influenced by Simondon, calls 'constantly inventing and re-inventing' oneself[24] and one's technical objects,[25] and, more generally, with process thinking and relational thinking. To move on, we have to replace those dualist and closed pictures of humans, their selves, their world, and their future with a process view of digital existence, in which selves, worlds, and meanings emerge from and are co-constituted in, processes of mutual and relational becoming. For Whitehead, the self is the result of a flow of relations. This is perhaps clearer when it comes to the growth and development of children. As Robert Mesle puts it: children 'must create their souls out of the relationships they find themselves in'.[26] But this process of self-becoming as a becoming-in-relation is never finished and, if we want to further develop, we need to remain open to the world and the people around us. It is a creative and relational ontogenesis in which we participate and to which we contribute. Both humans and their digital technologies are part of this ontogenesis, this becoming.

---

[22] Kreps, *Against Nature*, 59. Refers to Zuboff, *Surveillance Capitalism*.

[23] An example of transhumanist thinking can be found in Tegmark, *Life 3.0*. Other well-known proponents of transhumanism and Singulatarianism are Ray Kurzweil and Nick Bostrom.

[24] Schick, "Open Objects," 1.

[25] Following Simondon, Schick argues for 'open objects' that have 'an evolutionary potential' and that are not possessed but instigate participation and cooperation. Schick, "Open Objects."

[26] Mesle, *Process-Relational Philosophy*, 53.

This participative, process-oriented view of humans and digital technologies does not exclude responsibility, on the contrary. Normatively speaking, embracing a process approach implies that we must take responsibility for ourselves and our future, rather than hide between technologies such as AI and absolutize them—a practice of which Schick accuses transhumanists.[27] It means that we take on our task to participate in processes of what Bergson called creative evolution[28] and that we face digital existence and the adventure of life without expecting technology to solve our problems. It also means facing the future as an open invitation and accepting the uncertainty, risks, vulnerabilities that come with that. It has already been argued that we better acknowledge and accept uncertainty, against the denials and quick (technological) fixes. Uncertainty enables creativity and 'brings with it possibilities'.[29] It opens up the horizon of the future. Erinn Gilson claims that vulnerability is not necessarily negative.[30] And I have proposed a philosophical anthropology of vulnerability, which sees being vulnerable as part of human existence.[31] A process philosophical approach can support such efforts towards a more adequate description of human existence and what I have called 'normative anthropology'[32] and help us to see the normative implications. If the technological future that transhumanism presents us does not exist, if the future is fundamentally uncertain, then what remains is our nakedness within, participation in, and responsibility for, a world that is continuously in the making and that is in principle *unpredictable and unfixable*. It is within such a world that we must engage, relate, and respond. As Akama, Pink, and Sumartojo put it: 'Embracing uncertainty involves acknowledging that we do not and cannot know exactly what will happen next, and engaging with the possibilities that this affords'.[33] Such engagement is our ethical-existential responsibility, understood as a

---

[27] Schick, "Open Objects."

[28] Bergson, *Creative Evolution*.

[29] Akama, Pink and Sumartojo, *Uncertainty and Possibility*.

[30] Gilson, *The Ethics of Vulnerability*.

[31] Coeckelbergh, *Human Being @ Risk*. Ibid.

[32] Akama, Pink and Sumartojo, *Uncertainty and Possibility*. See also Bert-Jaap Koops's review of the book, "A Normative Anthropology of Vulnerability."

[33] Akama, Pink and Sumartojo, *Uncertainty and Possibility*, 36.

relational responsibility for our vulnerable participation in processes with uncertain outcomes.

This position on taking responsibility for our existence can also be interpreted as a revision if not rejection of Sartrean (and Nietzschean) existentialism: one that lacks the harsh quasi-protestant absolute responsibility and absolute freedom that Sartre theorized and one that rejects or at least updates Nietzschean nihilism. First, it is not about absolute responsibility since we are always in relation. We do not act in isolation but always interact. We are part of processes. This implies that we are not *absolutely* responsible. Second, it is not nihilist in the sense that meaning is seen as emerging from processes, in which humans but also technologies take part. The world as process is not empty of meaning. But the position still insists on human responsibility. Today, long after Nietzsche proclaimed the death of God, there is not only no anthropomorphic divine entity to whom we can offload our responsibility and freedom; we also cannot make *technology* into a God to which we can delegate responsibility. Consider the narrative of AI as Singularity: such technologies and narratives cannot and should not take away our existential responsibility. Not only as persons but also as humanity, we have to face our existential freedom and grow up, that is, use our freedom to take responsibility. But understood with process philosophy, this means: use our freedom to take responsibility for becoming and co-becoming: with others and with technologies.

Existentialists see existential responsibility as a burden. They say that because they picture us as isolated individuals burdened with individual and absolute responsibility within a world devoid of meaning. But the good news is that the nihilist picture gets it wrong. If we are part of a becoming world that is in the making, then we also make, and participate, in meaning and meaning-making. The world is made meaningful in the process, and so is our existence. Moreover, we are not alone, and others are not just and not always hell (to use a famous phrase from Sartre). Beyond Sartrean existentialist and individualist conceptions of morality, we find not only morality in the form of demands that seem to come to us as external demands or individual virtues, for example the demands of Kantian morality or the 'unselfing' Iris Murdoch wrote about in response to Sartre and the moral philosophers of her time—morality is less about my freedom and will and more about *overcoming* the focus on the self by attending to the beauty of reality[34]—but also morality that takes form in the concrete social and communal life understood as processes of co-existence. Inspired by philosophers such as Aristotle, Dewey, and Levinas (but also for example feminist philosophers), we must acknowledge that unselfing is in the first place something that I am called to by others[35] and in relation to others. And responsibility is not only something *I* face but also something *we* face as a community and a society. The responsibility question arises as a social problem and within social and relational processes. This also holds when it comes to responsibility for technology. Together we can and must participate in the responsible development, use, and maintenance of technology. And we are not isolated selves but are part of the world, understood as process and co-existence. We participate in processes larger than ourselves, including social-technological processes. We co-evolve. We co-exist. The existential burden and the freedom that comes with being and becoming in the world can be shared and is shared, at least to some extent. For example, we can and must face climate change as a problem of local and global co-existence and a problem of co-evolution. And imagining and engaging with the possibilities we encounter under those conditions can and must be done together and must include thinking about our relation(s) to technologies. (In the last part of this book I will say more about the politics of co-existence.)

---

[34] In Iris Murdoch's writings, for example those collected in *Existentialists and Mystics*, we encounter the idea that becoming moral has to do with developing a 'loving respect for a reality other than oneself' (218). To illustrate this 'unselfing', Murdoch describes how she looks out of her window and sees a bird. This alters her state of mind: 'The brooding self with its hurt vanity has disappeared' (369). However, Murdoch's (quasi?)mystical and Platonic account of morality does insufficiently recognize the concrete social dimension of morality and its vision-centred Platonicism is not obviously congruent with a more performative and process philosophy way of thinking.

[35] In contrast to Murdoch, who seems to derive ethics from metaphysics, Emmanuel Levinas argued that ethics, and in particular the ethical relationship, comes first. It is the other who requires me to respond. See Levinas, *Totality and Infinity*.

Narrative

Our digital existence and self, understood in terms of events and processes, has a structure. This is a temporal structure, of course, since we live in time and are 'lived' *by* time (for example when we are *being* hurried in a digital culture of speed and acceleration), but it is also a social and *narrative* structure. The temporality of human existence and its social aspect, co-existence, are structured in a narrative way. The point is not only that we humans are storytellers, as Yuval Harari[36] and many others recognize, but that we *are* a story. Our existence, including our digital existence, is a story. It has the structure of a story. And meaning may emerge in and from that story.

This thought is not new and can be found, among other places, in the philosophical tradition of hermeneutics. According to Paul Ricoeur, human life and experience have a narrative structure. His starting points are ancient Greek tragedy and text. In *Time and Narrative*,[37] he draws on Aristotle's *Poetics* to present a theory of emplotment and mimesis: the plot of a story organizers characters, motivations, and events into a meaningful whole. The narrative as a whole then makes sense and then leads to a new understanding. Like process philosophy and existentialism, this narrative approach gives us a non-essentialist understanding of the self. Here we understand ourselves and our existence through narrative, which means that we are not a fixed, given thing, but instead develop and become. And through narrativity we create meaning. Via narration, we draw together past events into a meaningful whole; we thus create and interpret our self, and understand the potentialities for the future. Digital technologies, then, support and shape this process of self-making, self-interpretation, and self-understanding. Today we make sense of ourselves through social media, for example. Digital technologies and media are thus not just tools to do particular things, say to search for information or to stay in touch with a friend, but also at the same time help us in making ourselves and letting ourselves emerge and co-exist. Technologies and media are thus contributors to narrative processes of becoming.

---

[36] Harari, *Homo Deus.*
[37] Ricoeur, *Time and Narrative.*

Combined with process philosophy, narrative theory also enables us to talk about the meanings that emerge from digital processes as a whole and describe the entire process in a way that goes beyond subject/object dualism. When the temporality of the digital processes and events is structured in a narrative way, meaning may emerge in a way that draws together nature and culture, humans and non-humans. A particular narrative (e.g., about climate crisis) structures characters (e.g., a particular animal threatened by climate change, specific scientific data about climate change) and events (mediatized and scientifically studied climate events) into a coherent whole that shapes the meaning of the whole (e.g., what we now call the climate crisis). The crisis is 'made' in a hermeneutical process in which we interpret and narrate. Since it is part of that process, but it is not the whole story. Climate change is not only measured but also narrated. One could say that climate change is at the same time science and a story. It is about facts and numbers; but it is also at the same time a process of interpretation and narration. The same can be said about pandemic and war. In such processes and narratives, science and culture, technologies and humans, objects and subjects are entangled. In contrast to common ways of (modern) thinking, which try to keep subjects and objects separate, process thinking offers a non-dualist way of conceptualizing climate change as process and narrative and can make sense of it; so-called subjective and objective elements come together and are made in processes, events, and narratives.

The narrative approach also enables us to highlight the social dimension of digital processes and events, and indeed the social dimension of existence. Narratives are not just personal—say, the narrative of my life time—but also involve others: there are also other characters in the story and those personal narratives are linked to wider meanings in the community, cultural, and society, meanings which may also take a narrative form. For example, what I do about climate change is influenced by what friends say and do, and in my culture there are already apocalyptic and eschatological narratives available, which influence narratives about climate change in my culture. These narratives are then integrated in my personal stories and shape them. For example, a climate narrative such as that presented by the Extinction Rebellion movement suggests that we move towards an 'end time' in which (part of) humanity becomes extinct.

A transhumanist narrative predicts that we move towards a Singularity or another end and destiny of humanity. These narratives then may be interpreted and integrated in personal lives and stories. For example, I may come to see my story as one that is about fighting for the survival of humankind or about promoting the development of superintelligent artificial agents that may replace us. Again we get a more relational approach, this time by using the concept of narrativity.

Moreover, inspired by Ricoeur, one can also take a hermeneutic and narrative approach to understanding digital technologies[38] in a way that shows how we are shaped *by* narrative and by technologies (and not just use them and create stories *about* them). As Wessel Reijers and I have argued, digital technologies play an active role in making meaning, in particular by configuring our narratives.[39] One could also say: technologies co-author our narratives. For example, a narrative about the end of our civilization because of climate change is not only made by humans that tell a story but also by the scientific instruments and the data collecting and data processing technologies used. These technologies thus have an active narrating role; they become a kind of co-authors of the narrative. At the same time, these technologies emerge as climate technologies from the process. They also become part of the story; they become characters in the story that get involved in events and that interact with other characters such as scientists, climate activists, and politicians. Like Bruno Latour (see also below), we arrive at a picture of the social that includes humans and non-humans,[40] but unlike Latour, we can highlight and better conceptualize the temporal and narrative dimension, and we are able to talk about meaning and existence.

Structured in a narrative way, digital processes and events thus lead to new subjects and objects, and also to new meanings. This reflects back on existential temporality, on our relation to time. The technological processes and events and the meanings that emerge from them through narrativity also shape the meaning of our time, our existence, and our selves.

---

[38] Romele. *Digital Hermeneutics*. For earlier work on Ricoeur and the philosophy of technology in general, see Kaplan, "Paul Ricoeur."

[39] Reijers and Coeckelbergh. *Narrative and Technology Ethics*.

[40] Latour, *We Have Never Been Modern*.

For example, a particular climate change narrative (for example an apoca-lyptical one), supported by digital technologies and media, shapes what it means to live on this planet today. It shapes our time and existence. It shapes *my* time and *our* time: my life time, our time in this society, and our time on this planet. Even if I may not intend it, I become part of such a climate change story. I become part of the narrative. Through these nar-ratives and processes of interpretation and meaning making, I become not only a climate crisis subject but also, or more precisely, a character in a story. For example, through a story about climate change as presented by classical and digital social media, I may become the bad consumer who contributes to climate change by eating meat or the heroic teenager fighting against climate change.

My entanglement in such narratives does not mean that I have to accept the narrative and its outcome. If I do not like the narrative and my character, I can try to change it. A narrative approach is relational and communicative. As participant in narrative and communicative pro-cesses, I can question the narratives of my community or the narratives I find on digital media. I can propose a different interpretation and discuss that interpretation with others. I can try to tell a different story: about myself, about others, about my society, about climate change, and so on. I remain responsible: for the narrative I tell and live, and towards others. I have what I have called *narrative responsibility*.[41] Furthermore, also in terms of temporality, narrativity always already has a social and collective aspect: the narrative connects my time to your time, and my time to *our* time. The narrative organizes us, our technologies, and our events. It shapes co-existence and gives meaning to it. It makes sense of what oth-erwise would be unconnected things and events.

This is true for the past and the present, but there is also the orienta-tion towards the future. Here making sense is linked to the normative. Narratives makes sense of the past but also help us to define our common future. They help us to imagine preferred outcomes for which we then strive, thus not only assisting us to make sense but also making the world as it unfolds. Narratives may coordinate us, for example to take measures to deal with what has emerged as 'the climate crisis', which then leads to

---

[41] Coeckelbergh, "Narrative Responsibility and AI."

a different world and a different future. And we are invited to continue or change the narrative, to create different characters and events, to generate new meanings. The hermeneutical process is never finished. Meaning emerges but can and will change again. Stories will change. There will be new stories, new meanings, new communications, and in the end new communities and collectives, new forms of co-existence. Process philosophy helped us already to set the picture in motion; narrative theory makes us understand at least one way in which processes are structured and how meaning is generated in a social context—which has in turn again a process character. Moreover, narrative theory further contributes to thinking about responsibility as narrative responsibility in a social context and in social processes.

Performance

But the conceptual work is not finished. While process philosophy and hermeneutics give us some concepts to talk about the experience and meaning of our digital existence, in particular about how that existence emerges from processes and narratives, we also need to further conceptualize what we humans and technologies *do* in these processes and narratives (apart from interpreting and narrating), and how this involves the *body* and concrete bodily movement. Understood superficially, classic process philosophy might offer a naturalistic account of what we do, for example in relation to climate change (how we contribute to global warming), but remain rather abstract when it comes to specifying the more concrete existential meaning of this and the implications for what we do in our daily lives. And narrative theory à la Ricoeur can frame our actions as the actions of a character in a story but is very much focused on language and—at least without further development—risks to fail to connect to views of human being that focus on bodies and movement and to hinder moving towards a more political understanding of what we do with digital technologies. In order to fully understand digital time and existence, therefore, we also need to say more about bodies and about what *humans* do (to each other).

In response to these shortcomings, I propose that we use the concept of *performance*, which is able to combine some of the insights from the other theories while helping to bridge the mentioned gaps. First, performance is also a process and may be governed by a particular narrative. But understood as a metaphor borrowed from the arts (for example dance or music), the term explicitly brings in the social and bodily dimension. This helps us to think about what we do with technology[42] and to further develop the process-oriented and narrative approach. We can still say that subjects and objects are produced and emerge in digital-technological processes and from digital-technological narratives, but if we see these processes as performances, then we can further highlight how our dealings with digital technologies also involve acting with others (an audience, co-performers) *and* how this always involves bodies.

Ricoeur's theory is based on ancient drama theory, and it focuses on the medium of language, on text. But there is also life and movement in and outside the theatre. People do not only speak but also move. They use language but also experience their body and move with their body. The temporality of my existence (or of specific chunks of my existence, a particular event for example) should not be seen as an abstract process or only in terms of narrative, as if there was a disembodied self or 'I' that is part of abstract narratives and processes. Instead, my time and my life time are always already both social and corporeal. Time and temporal processes do not make me in the sense of a disembodied mind or self; time also shapes my body and my relations with others. My existence is always both embodied and social. For example, my time in the light of climate change, and indeed climate change itself, are also a matter of me as an embodied person. My body and its vulnerabilities are connected to wider global and climatological processes and events. My body may be harmed by an extreme event, for example, and the way I sustain my body (e.g., with specific kinds of foods) may involve harm to the environment and contribute to climate change. Narratives and processes related to the pandemic organize not just an abstract 'me' but also me as a moving person and as a person with a body: it choreographs how I move in the city (for example walk rather than take public transport), what I do when I

---

[42] Coeckelbergh, *Moved by Machines.*

enter a store (put on my mask), how I pay (avoidance of touch), and so on, and it affects my body, for example when I get ill but also when I have stress because of the pandemic. The same can be said about narratives and processes of war: what digital media tell me influences my body and mind. The world changes and therefore I change. Things move in the world and I move (differently). The effects of media are also bodily and kinetic. My technologically shaped becoming, for example in the light of climate change, pandemic, and war, is an embodied and moving becoming. I am moved, halted, and twisted by these events, processes, and stories. To exist as process and in process also means to move and to be moved.

However, I also participate in these events, processes, and stories and thereby co-shape them. I am not just passive with regard to these processes and stories; me and my life (time) and our lives and (life) times are not just the passive outcome of these processes and stories; we and I also *perform* climate change in the sense that we talk about it and do other things that might influence processes and events. We and I also *perform* the pandemic and the war in this sense. This is only partly a matter of language. As we know since the works of J.L. Austin, John Searle, and Judith Butler,[43] words do things. Language has effects, and we use language to attain particular effects. Language can be used in a performative way. How I talk about climate change or the pandemic also shapes its meaning. It is not *just* a matter of science or of what we do otherwise (that is, without using language). What climate change 'is' depends on what it *becomes* in performative and communicative processes by scientists, politicians, citizens, activists, and so on. There is no unmediated epistemic access to what climate change 'is'. But performance also means bodily movement and performing with others. Performances in protests and climate activism performances, for example, co-shape the meaning of climate change in a way that involves movements with and of bodies. For example, performances by the Extinction Rebellion movement and by Greta Thunberg co-create the interpretation and meaning of climate change as climate 'crisis', with its specific temporal and quasi-eschatological meaning, and this involves bodily movements. Similarly, linguistic and

---

[43] Austin, *How to do Things with Words*; Searle, *Social Reality*; Butler, "Performative Acts."

bodily communicative performances by experts, politicians, and citizens co-create the meaning of the pandemic, of the war, and so on.

These performances are often mediated by digital technologies. Next to language use and use of the body, use of digital technologies also shapes those performances. To the extent that they do so, *digital technologies and media are also performative*. Consider for instance the role of social media and classic media such as TV in shaping the climate 'crisis' and the sense of urgency—and hence climate-related existential temporality. The same can be said about for instance the COVID-19 pandemic and the war in Ukraine. The technologies and media co-shape (what it means to live in) pandemic times and war times.

This means that while I actively perform, there is also a passive aspect. I perform, but I am also part of performances that happen outside my control. Participation and performance does not mean full control. Digital life and existence are not just a matter of processes that I control as a human being. Things also happen to me, and technologies co-shape what I do and what we do, including how I speak and how I move. As I have proposed in *Moved by Machines* and related work,[44] the term performance can be used to conceptualize our dealings with technologies, including digital technologies. As we use digital technologies and are involved in *technoperformances*, we move and are choreographed, we act and are directed. To conceptualize the important role technologies and media play, we can say that we do not only use technologies in order to move, but that technologies also move us and direct us. The medium is not only the message but also the performance. Technologies and media organize us, and thereby also organize our temporalities and existence. *Technologies and media shape how we perform time* and how we exist. We are participants in *technoperformances of time*.

I use the term *technoperformances of time* rather than just 'technoperformances' since, if process philosophy gets the nature of time right, it is not just my relation to time but *time itself that is performed* via technologies and media in these cases. When I use social media, this is not only a performance with technology but also a performance that shapes my time and, in the end, my life time and existence. In those performances

---

[44] Coeckelbergh, "'Technoperformances'.

and processes, specific subjective existential and scientific objective times and time lines come together in one process and performance. For these processes and performances, which create meaning but also make time and the world as it unfolds, we have narrative and performative responsibility.

Yet going against the abstract language of process metaphysics and in spite of the language of 'virtuality' that is often used in connection to digital media, these technoperformances of time are not an abstract matter at all. Digital processes and narratives and their temporalities involve both technologies and very real moving *bodies*, including the temporalities of those movements and those bodies. Our relation to time in digital processes and events is not just a matter of 'process' and 'narrative' in the abstract, but take a very concrete social, material, bodily, and kinetic form. The latter was already mentioned but also deserves emphasis in relation to time. When we use digital technologies, we literally *move* through time and we *are moved* through time. For example, I move with the cursor on the screen, I move through my emails, I scroll, I *move* through social media posts, and this involves use of my body (e.g., the *bodily movement* of my hand, arm, and eyes) and of digital technologies and their material form and material infrastructures, which in turn are in flux and require embodied humans and their movements to maintain them. I also use my mind, which is again related to my body and my movements since my thinking is embodied.[45] And when I am online, the offline movements and temporalities of my body do not stop. When I perform on and via digital social media, I do not leave my moving and vulnerable body behind, and neither do I get rid of my embodied mind. I perform with media and technologies as a performer but also as a fully embodied being-in-becoming. As a user of the technologies and media, I become as an embodied *becomer*, an embodied *emerger*.[46] As embodied becomers and emergers, we move through time and are moved by time.

To conceptualize these performances in terms of embodied and kinetic becoming means again that I am not in full control of what emerges in

---

[45] See the tradition of embodied mind and enactivism in cognitive science and philosophy.

[46] The term I propose here seems to be in line with the interpretation Kreps develops in *Bergson, Complexity and Creative Emergence*, which connects Bergson to post-Darwinian biology.

and from these performances. The narratives I perform tend to be only partly written by me. I am an embodied, social, and interactive *participant* in these technoperformances that organize temporality and existence, and from which temporality and existence emerges. My performed actions and movements are situated in time and are shaped by the relevant temporalities that are already available in my body and in its social-kinetic environments, as much as our performances themselves also shape time and temporality. For example, as I live and move on this planet, there are already climate temporalities and temporalities of climate politics in which I find myself and to which I may contribute through my performances. The pandemic and the war also offer their own temporalities, which influence my performances and only become reality through the performances of people and their bodies. Consider for example the temporality of how a particular Corona virus disease develops or the temporality created by the movements of an army. Both technologies and others play an important role in this 'making of time'. The way others move influences whether I become ill or whether I get killed. In these existential performative contexts, movements and time are a matter of life and death. I also perform and this also has some influence (for example words or bodily movements that are meant to be performances of resistance), but these performances are participating in processes, stories, and performances that are larger than what I do. My becoming(s) and performance(s) participate in larger becomings and performances. Together with others and with technologies, I am a participant in temporal existential processes, a co-writer of existence-shaping stories, and a co-performer of time. In this process-narrative-performative way, I become and emerge. Nevertheless, I remain co-responsible for my technoperformances and my time-makings, I remain responsible for my becoming(s).

The social dimension of technoperformances of time also implies that they are *political* processes: processes in which concrete human beings do things together and to one another—in time, in a specific rhythm, in a specific period, and so on—and in which technology and *others through technology* exercise power over them. The power dimension is connected to the relational aspect of process, narrative, and performance. For example, I can only contribute to the climate story that others co-write, whereby some are more powerful writers than others and their story dominates social media. I can only do *my* performances in the pandemic, while my existence-as-performance depends on the performances of others. Moreover, the performance of some (for example unvaccinated people having a party, an army that moves forward) has more influence than other performances and, in turn, influences my performances.

Technology and media play a role in this power-full process, in which technoperformances become political. This political aspect can be obvious. For example, a TV item about the Corona virus influences in turn my performances in the city. The rhythm of war also influences my time and experience, for example when I scroll for news items about it when I wake up. How I spend my time in everyday life is shaped by what others do and the related temporal structures that go beyond those of my personal life. But there are also more subtle ways in which technoperformances exercise power over me and my time, for example through electronic calendars, social media posts, emails, and digital messages. This is also 'political', in the sense that it is about power. (I will say more about this in the next section.) Our technoperformances of time are thus governed and choreographed by narrative and performative structures and other temporal structures at various levels, which are politically relevant and which shape our temporality, existence, and co-existence through technology. For example, by influencing the relevant narratives and performances, digital social media—and others *through* these media—co-shape what the climate crisis is and means, what pandemic means, and what war means. They may also (try to) co-influence and shape my daily time and performances, for example my ways of traveling (public transportation rather than car, train rather than airplane), all of which amount to significant changes in how I relate to time in my daily life.

This emphasis on the role of technologies and media does not diminish the role of the human. Our technologies and media can only have this power *through* human performances: directly through our own performances and indirectly through those who want us to do particular performances (rather than others). Consider for example a calendar which organizes us into a meeting, which is a performance with bodily and temporal aspect. Even in an online meeting, we still perform as embodied becomers and embodied minds-in-process. Now one could say that in such cases, the calendar governs our performance and movements; the relevant temporalities are shaped by the calendar and through our performances. In that sense, we are governed by the temporalities of the technology. But this only works if (1) someone performs in a way that changes our schedule, that is, if someone makes us do this through shaping the digital calendar via performative acts such as use of words (e.g., the technoperformative act of an email that has already a zoom link in it) and through collective, that is human, use of a shared calendar, and (2) if we then also perform based on the calendar, if we follow its temporal-performative structuring in and with our performances: if we actually meet, if we move to the meeting, if we perform according to our digital calendar. In other words, not only through technology but also through these human participations power is exercised. Power is performed: on me and by me, albeit mediated by technology. The technology and medium enable others to co-shape my time but all this also depends on my performances.

This human participation in technoperformances enables us to retain our responsibility for them and also makes resistance is possible. If there was technological and social determinism, this would not be the case. But since human participation is required in technoperformances, resistance is possible in principle. In practice this may be difficult, for example when the technology, that is, the technology of the calendar, makes resistance harder: what others want us to do is already embedded and incorporated in the calendar and the related technoperformances. As such, the calendar supports and encourages a particular structuring of (our technoperformances of) time. To go against that requires more effort and courage—and perhaps also other power-full technoperformances, counter-technoperformances. If we do not like how our performances are governed and structured, by others and by technology, we have to actively make time in a different way. We have to actively resist how our time is governed, directed, and choreographed. Technologies and media, in the form of technoperformances in which we participate, render this more difficult as they pre-structure our time and thus shape our temporality and daily existence. Technoperformances of time thus involve, establish, and maintain relations of power, in which both humans and technology play an important role. And this has existential and societal consequences.

Not only my work life but also my life and life *time* (not just life world) as a whole are shaped by these technoperformances. This has political significance, since it is about who shapes my time, my life, and my existence.

That and how others shape my time is not always clear when we focus on the (use of) technologies and media. Technology appears in these performances as a kind of actor, whereas the human actors and stakeholders are hidden from view. I therefore might experience the *technology* as doing something to me.

That experience is partly right: technologies are co-performers and structure the temporality of my performances. Digital technologies, which enable some degree of automation, are very power-full in this sense. I experience that I am moved by my calendar, which tells me I need to move to another meeting. And the influence extends to my life and existence as a whole, which under conditions of a wider social and narrative context becomes hurried, anxious, and tiring. Calendars and phones respect no boundaries.[47] After the end of my work day, I move into my free time but my smartphone encourages me to be available. Through the technology, I am expected to be ready to answer, to move quickly, to perform—also in the sense of delivering. The office is everywhere. I move through my life and I *am* moved through my life in ways that are shaped by the digital technologies I interact with. In principle I am free to put away the smartphone. But the smartphone as a technology and medium discourages me to do so, and through the smartphone another specific technology, the digital calendar, exercises power over me. One type of time performances is encouraged, whereas other types of time performances are discouraged. I am encouraged to move in these ways rather than others.

Those movements interact with, and are influenced by, other movements: movements in my body, movements in nature and the universe, and movements by other people. Consider 'biological clocks' but also many other 'clocks' that move and in turn make us move. Again, digital

---

[47] Using the methodology of sociologist Eviatar Zerubavel, Elpida Prasopoulou and colleagues have analyzed how mobile phones disrupt the boundaries that separate different temporal spaces. Prasopoulou, Pouloudi, and Panteli, 'Enacting New Temporal Boundaries'.

technologies are sometimes interpreted as if they no longer involve the body or bodily movement, but we always move when we use the technologies and *are* moved in specific ways by the technologies. For example, a designer has choreographed the movements I have to make on my smartphone to use a specific operating system and a specific app. If I want to operate the phone, I *have* to move in those ways rather than others. The designer has thus choreographed my technoperformance. I also follow the rhythm of a specific medium, for instance the tantalizing rhythm digital social media (each of which also have their own rhythm) rather than, say, the scientifically mediated slow rhythm of geological change and historical climate change that is read from the growth rings of trees or from analysis of air trapped in ice from ancient times. Scientific technoperformances produce this knowledge and at the same time also potentially shape a temporal structure for my own technoperformances, for example my technoperformances in the supermarket when I buy climate-friendly food. The medium is the message, and in the form of technoperformances could conclude that the medium is my time and the medium is my existence.

Technology thus plays an important role in shaping time, including shaping my time and our time. But this should not be an excuse for evading responsibility and for neglecting the political dimension. Through the participation of humans, technoperformances are and remain both human and political, and this gives us a responsibility: as humans and persons we are at least partly responsible for our technoperformances, the narratives by which they are structured, and the temporalities and existence(s) that emerges from them, ours and those of others. Process thinking, narrative thinking, and performance-oriented thinking thus enable us to conceptualize both the role of technology and the role of humans in time making, and provide a basis for a normative discussion about responsibility in the light of these processes, narratives, and performances.

Further Discussion: Multitemporality and Power

Taken together, the proposed concepts (process, narrative, performance) enable a less dualistic way of thinking about digital existence. They show that for thinking about digital time and existence, it is neither necessary nor desirable to stage humans against technology or subjective against objective time. Such tensions may be experienced, of course. They are part of the phenomenology of everyday life. Our experiences are still largely governed by dualist, non-process modern thinking. Modern thinking continues to be experienced, enacted, and performed. But at a more fundamental level, digital experiences, processes, and performances themselves must be understood in a non-dualist way as being about both subjects and objects and as being deeply relational. The tensions themselves are not ontologically primary but are produced in processes, narratives, and performances of meaning making through digital technologies. They are themselves emergent from particular ways in which subjects and objects are constituted in, and emerge from, technoperformative and narrative processes. Ultimately, time itself—or more precisely, particular times—emerge from these processes. Time is not given but it is made and emerges from process, including technological process. It *is* also that process. For example, in the case of the digital calendar, technology and time cannot be separated. Adding the concepts of narrative and performance to process thinking thus helps to further develop process thinking about time, technology, and existence.

Moreover, my use of these concepts is in line with accounts that stress the plurality of time, which I now propose to conceptualize in terms of a plurality of technoperformative processes, from which a plurality of temporalities may emerge and which may combine different temporalities.

Let me unpack this. It has rightly been argued that there are different times, *temporalities* (plural), and speeds. For example, Andrian MacKenzie has argued that there are differences in speed, rather than a general acceleration of everything. We feel differences of speed, rather than speed as such.[48] Judy Wajcman and Nigel Dodd have asked whether acceleration differs between groups: some groups may be 'able to mobilize speed as a resource while others are marginalized and excluded'.[49] And as Barker observes, multiple times and rhythms are produced when we interact with digital systems. Digital temporality is 'thick' when it involves many dispersed moments, for example in the context of digital social media.[50] Moreover, data from the past become 'synthesized' in the present.[51] The past is made present again in particular ways.

Now Barker uses a process philosophy approach, but this can be combined with a performance-oriented approach to show how different performances create different times or bring these times together. For example, using digital media to learn about climate change and the Anthropocene brings geological and climate times together with life times, societal times, and civilization times. Pandemic and war are also producing different times, for example my clock-shaped day and the temporalities of disease development or of the advance of an army. In digital processes and performances, a plurality of times and forms of what we may call *thick temporality* may be created. My digital existence in the light of climate change takes place in the now but also in the future. Perhaps there are many futures, many scenarios. Consider the plurality of relevant pandemic scenarios or war futures. All come together in my present interaction with the technology, in my day-time, in our life time, and in our age. Similarly, AI and data science bring the past to the present, may re-*present* the past (and perhaps predict the future). And if my technoperformances take place in the hot and thick now of social media, then they might involve a synthesis (synchronicity?) of many 'nows'. For example, there is the thick hot present of war as *presented* by social media and as emerging from my scrolling and clicking performances, which is a kind of powerful black hole to which both histories and future scenarios gravitate. What the war 'is', is not only about what happens in the present but also a matter of interpreting the past and imagining the future, and

---

[48] MacKenzie, *Transductions*, 122.
[49] Wajcman and Dodd, *The Sociology of Speed*, 1–2.
[50] Barker, *Time and the Digital*, 29.
[51] Barker, 19.

through technoperformances (e.g., people posting opinions about the war) all come together in the technologically shaped 'now'. Shaped by digital media and indeed performing through various digital media and being choreographed and conducted by various digital media, we live in a hotspot of many presents, which all meet in our times and time-performances with the technology and our times and time-performances with others. We are perhaps not really multitasking when we interact with digital media, but for sure our technoperformances in and through digital media are multi*temporal*: in such performances, multiple temporalities meet. This may be at least *one* reason why digital existence is so exhausting and exciting at the same time: as we interact and perform with these technologies and media, we do not only live in several worlds but also in different *times*.

Note, however, that talking about a plurality of digital times does not mean that there is a separate 'digital time' as opposed to an analogue time. Here too, we have to maintain a non-dualist view. Just as there is no separate cyberspace as opposed to the spaces of lifeworld, there is no separate digital time or 'cybertime' as opposed to the times of the lifeworld. We live *in* digital times. There is a plurality of spaces and a plurality of times, but they are connected. With a process-oriented approach, we can say that both the digital and the non-digital are part of the same process, the same becoming. To think of 'cybertime' as a time separate from other times is thinking about time in spatial and dualistic terms. Instead, different times meet and are connected within processes, narratives, and performances. These cross and combine digital and analogue, online and offline. There is neither a separate lifeworld nor a separate 'lifetime'; both are already transformed by the digital and the digital is at once a life space and a life time. In this sense, digital existence, understood as process, narrative, and performance, is conceptually unified through process thinking about time. At the same time, the proposed framework enables us to recognize the plurality of interconnecting temporalities within these processes, narratives, and performances.

Furthermore, what I said about the politics and power dimension of technoperformances about time can be linked to the work of Michel Foucault. He showed how power is not just something that happens in the relationship between citizens and nation states, but pervades societal institutions and all relationships between people.[52] I already suggested that technoperformances of time also produce relations of power. For example, they invest some people with the power to organize our time and to decide about how to deal with climate change. With Foucault we can understand this in terms of disciplining. People may use technoper-formances (e.g., via social media) to discipline others to change their behavior, for example in the light of climate change, pandemic, or war. Moreover, we can use Foucault's term 'biopower' to talk about how bod-ies are moved and analyzed by these technoperformances. Digital tech-nologies are used to exercise biopower over pandemic subjects, for example via the collection, analysis, and use of data about (new) cases and deaths. And these technologies can be used to move bodies in a war or in a climate crisis. Finally, Foucault's later work on technologies of the self can be used to theorize how time is not an abstract concept but affects 'bodies and souls, thoughts, conduct, and way of being',[53] which again means that time is used to exercise power, but also that time at the same time subjects us and makes us into (particular) subjects. Again this is helpful to stress the existential impact of digital technologies. Digital technologies and the technoperformances that they enable are not just tools or technological actions, but shape our bodies and souls and our way of being. They are time technologies but also, through our perfor-mances with the technology, technologies of the self. For example, my 'body and soul' may be influenced by my regular use of my smartphone to read about the pandemic or about the war. I may become more anx-ious, stressed, and so on. I then may try to self-discipline, for example by using digital technologies such as a running app or a meditation app.

Another source for understanding the political dimension of techno-performances of time are again the performative approaches proposed by Austin, Searle, and Butler. They taught that language is not just about what is the case but also about doing something, about changing the world. This is also applicable to *technologies* and technoperformances— and for the relation to time and existence that emerges from these tech-noperformance processes and events. It has become clear that, why, and

---

[52] Foucault, *Power/Knowledge.*
[53] Foucault, *Technologies of the Self.*

how technologies and technoperformance are performative. Neither processes and performances, nor the existential meanings and temporalities that emerge from them, are normatively neutral. Technoperformances (of time) do something to us and shape the world. Time itself, in the form of processes and performances, is performative. Technoperformances of time are also exercises of power, and they are ethically and politically relevant. For example, my use of social media to read and post about 'the climate crisis' involves political narratives and meanings and is itself a politically relevant performance. Furthermore, there are also collective and public aspects to time making; there are collective technoperformances of time such as those regulated by clocks. Technoperformances organize our time and our relation to time. In this sense, too, technoperformances of time are political.

Now if technoperformances are indeed political and if my stories and life time—and indeed *our* stories and *our* lives—are shaped by (digital-) technological processes and events, and if I, or at least *we*, have some influence on these processes and events and some responsibility for them, as I suggested, then the question arises: How can we *evaluate* these technoperformances of time? What would be an ethics of technoperformances of time, and what kind of politics of technoperformances of time do we need? What is a good relation to time, given digital technologies? What is a good time, narrative, and existence at all? And is this a matter of my (life) time only or can and should we also discuss about common time and co-existence, especially in the light of shared local and especially global challenges such as climate change and the pandemic? Who shapes, and who do we allow to shape, our technoperformances of time? Indeed, who shapes, and who do we allow to shape, our existence and co-existence—in this life, in this society, and on this planet?

References

Akama, Yoko, Sarah Pink and Shanti Sumartojo, eds. *Uncertainty and Possibility.* New York: Routledge, 2018.
Anscombe, Elisabeth. "Modern Moral Philosophy." *Philosophy* 33, no. 124 (1958): 1-19.

Austin, J.L. *How to do Things with Words*. Oxford: Clarendon, 1962.

Barker, Timothy Scott. *Time and the Digital*. Lebanon, NH: Dartmouth College Press, 2012.

Bergson, Henri. *Creative Evolution*. New York: Random House Modern Library, 1944.

Bergson, Henri. *Duration and Simultaneity*. Translated by Leon Jacobson. Indianapolis: The Bobbs-Merrill Company, 1965.

Butler, Judith. "Performative Acts and Gender Constitution: An Essay in Phenomenology and Feminist Theory." *Theatre Journal* (1988): 519-531.

Coeckelbergh, Mark. *Human Being @ Risk*. Dordrecht: Springer, 2013.

Coeckelbergh, Mark. *Moved by Machines*. New York: Routledge, 2020a.

Coeckelbergh, Mark. "Technoperformances: Using Metaphors from the Performance Arts for a Postphenomenology and Posthermeneutics of Technology Use." *AI & Society* 35, no. 3 (2020b): 557-568.

Coeckelbergh, Mark. "Narrative Responsiblity and AI." *AI & Society* (2021). https://doi.org/10.1007/s00146-021-01375-x

Deleuze, Gilles, and Félix Guattari. *Mille Plateaux*. Paris: Les Editions de Minuit, 1980.

Foucault, Michel. *Power/Knowledge*. Translated by Colin Gordon and Leo Marshall. New York: Pantheon Books, 1980.

Foucault, Michel. *Technologies of the Self*. Edited by Luther H. Martin, Huck Gutman, and Patrick H. Hutton. Amherst, MA: The University of Massachusetts Press, 1988.

Gilson, Erinn. *The Ethics of Vulnerability*. New York: Routledge, 2016.

Harari, Yuval. *Homo Deus*. London: Penguin, 2015.

Harman, Graham. *Object-Oriented Ontology: A New Theory of Everything*. London: Pelican Books, 2018.

Ihde, Don. *Technology and the Lifeworld*. Bloomington: Indiana University Press, 1990.

Kaplan, David M. "Paul Ricoeur and the Philosophy of Technology." *Journal of French and Francophone Philosophy* 16, no. 1/2 (2006).

Koops, Bert-Jaap. "A Normative Anthropology of Vulnerability." *Law, Innovation and Technology* 5, no. 2, (2013): 289-297.

Kreps, David. *Bergson, Complexity and Creative Emergence*. London: Palgrave, 2015.

Kreps, David. *Against Nature*. New York: Routledge, 2018.

Kreps, David. *Understanding Digital Events: Bergson, Whitehead, and the Experience of the Digital*. New York: Routledge, 2019.

Kreps, David, Franz Rowe, and Jessica Muirhead. "Understanding Digital Events: Process Philosophy and Causal Autonomy." In *Proceedings of the 53rd Hawaii International Conference on System Sciences*, 2020.

Latour, Bruno. *We Have Never Been Modern*. Cambridge, MA: Harvard University Press, 1993.

Levinas, Emmanuel. *Totality and Infinity*. Pittsburgh, PA: Duquesne University Press, 1969.

Lipscomb, Benjamin. *The Women Are Up to Something*. Oxford: Oxford University Press, 2021.

MacKenzie, Andrian. *Transductions*. London: Continuum, 2002.

MacKenzie, Donald. "Capital's Geodesic." In *The Sociology of Speed*, edited by Judy Wajcman and Nigel Dodd.

Mesle, C. Robert. *Process-Relational Philosophy*. West Conshohocken, PA: Templeton Foundation Press, 2008.

Murdoch, Iris. *Existentialists and Mystics*. London: Penguin, 1997.

Prasopoulou, Elpida, Athanasia Pouloudi and Niki Panteli. "Enacting New Temporal Boundaries: The Role of Mobile Phones." *European Journal of Information Systems* 15 (2006): 277-284.

Rambo, David Nathan. "Technics Before Time: Experiencing Rationalism and the Techno-Aesthetics of Speculation." PhD diss., Duke University, 2018.

Reijers, Wessel, and Mark Coeckelbergh. *Narrative and Technology Ethics*. New York: Palgrave, 2020.

Ricoeur, Paul. *Time and Narrative—Volume 1*. Translated by Kathleen McLaughlin and David Pellauer. Chicago: The University of Chicago, 1983.

Romele, Alberto. *Digital Hermeneutics: Philosophical Investigations in New Media and Technologies*. New York: Routledge, 2020.

Sartre, Jean-Paul. *Existentialism is a Humanism*. New Haven: Yale University Press, 2007.

Schick, Johannes F.M. "The Potency of Open Objects," *Techné* (2021) online first.

Searle, John R. *The Construction of Social Reality*. New York: The Free Press, 1995.

Shoshana Zuboff, *The Age of Surveillance Capitalism* (New York: PublicAffairs, 2019a).

Simondon, Gilbert. *On the Mode of Existence of Technical Objects*. Translated by Cecile Malaspina and John Rogove. Minneapolis: Univocal Press, 2017.

Tegmark, Max. *Life 3.0: Being Human in the Age of Artificial Intelligence*. London: Penguin Books, 2018.

Verbeek, Peter-Paul. *What Things Do*. University Park, Pennsylvania: Penn State University Press, 2005.

Wajcman, Judy, and Nigel Dodd, eds. *The Sociology of Speed*. Oxford: Oxford University Press, 2016.

Whitehead, Alfred North. *Process and Reality*. New York: The Free Press, 1978.

Zimmerli, Walther. "Künstliche Intelligenz und Postanaloges Menschsein." In *Künstliche Intelligenz - die große Verheißung*, edited by Anna Strasser, Wolfgang Sohst, Ralf Stapelfeldt, and Katja Stepec. Berlin: Xenomoi Verlag, 2021.

Zuboff, Shannon. *The Age of Surveillance Capitalism*. New York: PublicAffairs, 2019b.

3

In Search of Common Time in the *Anthropochrone*: Good Times, Contemporalization, and the Politics of Global Co-existence in Times of Climate Change

Abstract This chapter explores the ethics and politics of technoperformances of time. What are good processes, narratives, and technoperformances? What is a good time and meaningful existence? How can we find a common time and develop good co-existence? First the chapter considers ethics: moving beyond individualist interpretations of virtue ethics, it engages with Borgmann's concept of focal occasions and discusses Millet's painting *The Angelus* to explore what good times are and to argue for a non-romantic ethics of technoperformances of time, which always already disciplined. Today we need new technoperformances that create opportunities for co-existence and *kairos* to emerge, although it is not clear what this means in a global context, for example in the context of climate change. The example of a 'climate clock' is discussed. Then the chapter turns to the *politics* of technoperformances. It responds to literature about the politics of time to show that time making is political and asks how we can achieve more synchronicity, in particular a non-capitalist one. It is argued that politics is not only about finding common spaces but also about finding common times. The chapter also questions what it calls *anthropochronic* thinking: we should also respond to non-humans and their times and temporalities. To develop this idea, we can learn from other cultures, in particular indigenous cultures. Moreover, natural sciences such as geology can also help us to 'contemporalize' (in analogy to contextualize) our own time. We should look at the long term but avoid transhumanist longtermism. In the spirit of Arendt, the chapter ends with a call for new beginnings: we should assist the becoming of new dances, new forms of synchronicity. Anthropological work and new technologies can help with that. The challenge concerns finding and technoperforming the good times and the common times in times that are changing.

Keywords Romanticism • Borgmann • Kairopolitics • Synchronicity • Anthropochrone • Longtermism

Kairos: Good Times and Meaningful Existence

The previous part has helped us to understand time and existence in relation to digital technologies. What it means to live in digital times, with digital technologies and media, can be conceptualized by using the terms of process, narrative, and performance. Our digital existence is a matter of participating in, and emerging from, processes, narratives, and performances in which our digital media and technologies play a key role. It is a matter of becoming through digital technoperformances and the related processes and narratives. Yet I also pointed to the normative dimension of technoperformances of time. This raises further questions concerning the ethics and politics of digital existence and co-existence.

Let us start with ethics. What are *good* processes and narratives? What are good ways of performing with technology? What are good becomings? What are good ways to *perform time*? What is a good and meaningful digital existence? Such questions remind us about the ancient question about the good life, about *eudaimonia*, but now formulated as: what is 'good time'? For example, what is good time with digital technologies and in times of 'climate crisis' and pandemic?

An initial approach to the ethics of time and existence in the light of digital technologies, which connects to ancient ethics of the good life, could be to respond to claims about acceleration and anxiety with the recommendation that we need better humans and better persons. For example, the worry that we are speeding up things might be followed by a call for the virtue of patience. Shannon Vallor mentions the virtue of patience in her book on *Technology and the Virtues*.[1] Vallor's work is important since it presents us the challenge to reflect on what virtues we need in digital times. However, while there is nothing wrong with calling for more patience, for slowing down, and so on, virtue ethics is often interpreted in an individualist way, neglecting social and political issues (see the next section). Calling for people to change their individual character in a more virtuous direction but leaving wider social structures that work against this intact is problematic. For example, how to slow down when a capitalist order encourages us to further speed up and when digital technologies constantly push us to speed up?

Moreover, traditional approaches to technology ethics tend to assume a merely instrumental conception of technology, focus on the ethics of the human, and neglect the specific ways in which digital technologies, existence, and time are entangled—which I described in terms of processes, narratives, and technoperformances of time. Both contemporary philosophy of technology, for example postphenomenology, and the process approach to technology proposed here enable us to question the instrumentalist and dualist conception of technology, which separates humans and technologies, ends and means. Technologies are not ethically and politically neutral, but, in the form of techno-processes and technoperformances, shape our lives, societies, and existence. Our 'ends' emerge from our linguistic performances but also from our technoperformances and from the larger processes and narratives in which these performances are embedded. If we want to change our relation to time in digital times, then, it is not sufficient to ask humans to change. The entire process, narrative, and performance has to change, including the *technologies* that shape the processes, narratives, and performances that constitute our lives and existence. The questions to ask are then: How to change the process and in which direction? Which narratives are good? Can we make time and perform time in a different way, perhaps with different technologies?

One option is to use older technologies instead of digital technologies. Albert Borgmann proposed in *Technology and the Character of*

---

[1] Vallor, *Technology and the Virtues*.

Contemporary Life[2] that we should attend to *focal* things and practices. We should engage with things and interact with others, rather than using digital technologies and other modern technologies, which create distance and preclude a careful and meaningful relation with things and with others. In an article on cyberspace,[3] he proposes to embrace focal occasions: meaningful occasions such as a camp fire in the wilderness or an evening of music. He defines these as occasions on which the meaning of life comes into focus in the sense that one can affirm the following propositions:

1. There is no place I would rather be.
2. There is no one I would rather be with.
3. There is nothing I would rather be doing.
4. And this I will remember well.

Such moments, Borgmann argues, are related to what the ancient Greeks called the *kairos*: the auspicious moments. Focal occasions stand out in the rush of time, but are more regular than *kairos*. They are also communal. They are local. And they are, of course, focal. He contrasts focal moments to cyberspace, which he calls 'the great distraction'. The cell phone interrupts. There is no focal point. We need to make time for focal occasions, times that are 'secure from the intrusions of information devices'. Such devices should not be rejected altogether—Borgmann allows them to inform us—but they should not always be used. We need times without them, when we can have focal occasions and focal times. One could also say that without them one can have *good times*. In the vocabulary I proposed in the previous part of this book: if the technoperformances are good, good times emerge from them. For example, the technologies and techniques of camp fire making and music making enable technoperformances that lead to good times, or at least afford, increase the chance of having *kairo*-ethical moments. Furthermore, one may want to add to Borgmann's analysis that these moments constitute a genuine kind of present, which contrasts with the presentations and

---

[2] Borgmann, *Character of Contemporary Life.*
[3] Borgmann, "Cyberspace."

re-presentations by digital media, AI, and other digital technologies. In focal occasions, one is not distracted but present to the moment, to others, and to oneself.

Another way to achieve this is focusing on one's own body. Since ancient times,[4] one technique to get more 'focal' and mindful is focusing on one's breathing, thus re-connecting to one of the natural rhythms of the body, one of the body's key temporalities, and (re)turning to the present. It is a performance of present awareness. It is also a way of performing relationality: it establishes a flow between inside and outside, between me and the world. Digital technologies tend to divorce us from awareness of our relations to others and the world, from awareness of having a body and being a living and breathing being, and hence from awareness of ourselves and of what we call nature. They also tend to neglect or even negatively influence the human body and its functions, including breathing.[5] Digital technologies, in so far as they are performed in a dualist and alienating way, rather seem to enable the realization of the Platonic dream of leaving the body and the Cartesian exercise of retreating in the abstract 'I'. Having first conceptualized body and mind as things, they then perform a separation of both in a way that 'liberates' the mind. Transhumanists seek to perform the same when they entertain the idea of 'uploading' of the mind into a digital sphere. They attempt to cope with vulnerability at the cost of annihilating life itself. By contrast, focused breathing, alone or together, contributes to awareness of the very opposite idea: the affirmation of life as process and relationality, and the awareness of being one, being whole.

While these analyses make sense, they can easily become part of a romantic reaction against modern technology and a longing for a Garden of Eden[6] when times were still good. Borgmann's *kairo*-ethical criticism of the temporality of contemporary digital technologies stands in a longer tradition of arguments against modern time keeping. The idea is that even before digital technologies and the Internet, clock time already speeded up life and continues to tyrannize us.[7] In response, one may then

---

[4] Consider the history of Hinduism, Buddhism, yoga, and mindfulness.

[5] There seem to be measurable physical effects on breathing. An empirical study shows that prolonged smartphone use could negatively impact posture and respiratory function. Jung et al., "Smartphone Usage."

[6] This criticism of a Garden of Eden before technology is in line with Don Ihde's arguments.

[7] Zadeh, "The Tyranny of Time."

long for past times. And that longing is already part of Western modernity as shaped by Romanticism. We are in search of lost time, as the title of Marcel Proust's famous novel puts it.[8] We miss childhood, but we also long for meaning in the modern world. It is a romantic reaction against the nihilism and disenchantment already experienced in the beginning of the nineteenth century and later diagnosed by Nietzsche and (other) existentialists. In response, we may want to go back in time. But does one find the Garden there?

Consider the painting *The Angelus* by Jean-François Millet, which does not only express religious devotion but also re-presents a rural regime of time regulation and an associated performance of temporality.

The Angelus by Jean-François Millet, 1857–1859. Musée d'Orsay, Paris

---

[8] Proust, *In Search of Lost Time*.

The ringing of the bell and the performance of the praying ritual marks the end of the work day on the field. One might imagine a romantic rural form of temporality, long lost in today's world. But it is also a technoperformance, since the bell (which cannot be seen on the painting but is imagined, or rather *heard*, by the spectator) is a time technology that disrupts the work with other, agricultural tools. The bell structures the temporality of the peasants' work day. The way the field workers stand there shows that already in pre-industrial times, the time and bodily performances of people were regulated, yet without, perhaps, tyrannizing people. The painting shows us a different, non-modern technoperformance and *tempo-performance*. A focal one, if we borrow Borgmann's term, and also an embodiment of time. Time is incorporated and performed, and it seems that this form of temporality does not make the characters (or the spectator) sick; instead, there is a sense of healing that emanates from it. It is also a co-performance, a social ritual. There is harmony and communality. There is co-existence.

At the same time, the painting also reminds us that also in past times technologies played a key role in shaping time. In such agricultural societies people were also regulated by *technology*. Technologies and media have always shaped temporality, thereby changing our lives and thoughts: from Stonehenge and the Aztec calendar to today's computers and mobile phones and other electronic timekeepers.[9] And each time and in each age our tools have ordered and organized us in different ways. With a Foucauldian and Nietzschean twist, one could say that there was always *power*. Time (regulation) is power. These field workers and their performances are *timed and disciplined*, with the help of the bell. Those who discipline remain invisible. The church tower functions as a kind of Panopticon and without necessarily involving someone who watches: it enables a form of surveillance without being seen.[10] The regime is embedded in a techno-architectural and geographical structure, as well as in a religious narrative that structured people's lives and society. Later industrial factories and modern schools will introduce new forms of

---

[9] Birth, *Objects of Time*.

[10] I refer here to Jeremy Bentham's panopticon, an architecture used for prisons, hospitals, and other institutions, and discussed by Michel Foucault in *Discipline and Punish*.

disciplining and disciplining technologies, although sometimes old technologies were still used, for example the bell.

However, without being overly romantic, one may argue that while in rural contexts and pre-industrial times there is also disciplining and ordering, time and life is not yet fast. There is an opportunity for *kairos*, for creating a meaningful focal moment. There is also the possibility of a common sense, a common way of relating to the world. Like Eastern rituals and meditation practices, older Western rituals such as the Angelus and its associated technoperformances and narratives can create such occasions, although nothing is guaranteed. The meaning making here works with time technologies that regulate a cyclical rhythm (albeit human-created) rather than merely linear clock time.[11] Timing the end of the work day is meaningful. The emergence of meaning does not just come from the religious framework but is possible *because of the timing* and the performance, and technologies enable that. A painting such as Millet's shows us a technoperformance of time from the past and invites us to develop a hospitality for other forms of temporality, next to reminding us about the ethical importance, problems, and possibilities of technoperformances in general.

This interpretation need not be romantic. Yet romantic thinking remains seductive. And it seems that the acceleration narrative still makes sense. Whereas the painting makes us dream of past (lost?) forms of temporality, the modern world is ruled by the clock. We are living in the empire of *chronos*. Digital technologies seem to support clock time and its related temporalities and forms of existence. Time is pressing. Time is money. Time is lacking. In the context of environmental, health, and climate concerns, the point is usually that clock time is not in sync with our biological time and the time(s) of nature. Our existence seems out of joint or rather (to avoid a place metaphor) out of time. We exist in a state of temporal alienation.

This argument could also be applied to the climate crisis. We are put on a time line and presented with a deadline. The logic of the climate crisis is a chrono-logic. As Joe Zadeh remarks, the climate crisis itself is

---

[11] For more historical context to the distinction between cyclical time versus linear time see Gould, *Time's Arrrow, Time's Cycle*.

constantly translated into linear time by means of clock deadlines. He suggests that, paradoxically, this 'has contributed to the inability and inertia of many to comprehend the seriousness of what is actually happening' and has even produced the ecological crisis.[12] Ignoring nature's temporalities and imposing clock time on everyone and everything, we have created the Anthropocene[13] or rather what I propose to call the Anthropo*chrone*, which can be defined as the age of the human and the *time* of the human, in particular the chronological time. We have changed the planet into a space ship managed by modern clock time. We try to synchronize and coordinate humans and other animals by means of clock time, turning the whole planet into a gigantic clock and a time bomb.

Consider again Mumford's work but also Foucault, who wrote in *Discipline and Punish*[14] about time tables, which were first used in monasteries and later in disciplinary institutions. This disciplining is now applied to humans and non-humans. But we do not only discipline others and their time; we also try to self-discipline and self-domesticate and have created local and global chrono-performances and techno-performances in order to do this. Within this narrative, digital technologies turn out to be the instruments of a planetary scale anthropocentric and globalistic synchronization project against nature.[15] In terms of temporality, we have not only alienated ourselves from nature and made ourselves into temporal aliens on earth; we have also forced (the rest of) nature into our modern and human time regime—or at least we try.

In response, one might try to do without advanced time-keeping technologies and long for past times, as Millet already did in the nineteenth century, when industrial society was just emerging. One might try *to turn*

---

[12] Zadeh, "The Tyranny of Time."

[13] The Anthropocene is a controversial and unofficial name for the current geological age, which is the most recent period in the Earth's history and is seen as the period in which human activity becomes the dominant force on earth and significantly impacts even the climate and the environment. The term was coined by Eugene Stormer and Paul Crutzen. It is not clear when this period began. Some say the start of the industrial revolution; others think that it only started in the mid-twentieth century. The official name for the current geological chunk of time is the Holocene, which began after the last major ice age. Source: for an accessible summary see for example https://www.nationalgeographic.org/encyclopedia/anthropocene/

[14] Foucault, *Discipline and Punish*, 149.

[15] See also Kreps's argument that the philosophical basis of digital information systems as linked to positivist philosophy and capitalism is fundamentally against nature. Kreps, *Against Nature*.

back time (a metaphor based on the mechanical clock), thereby confirming rather than resisting the clock time way of thinking. The romantic response to acceleration remains entangled with the modern technological way of thinking. Both are stuck in a perpetual dualist dance.

Another (and in practice potentially compatible) direction is to reject the romantic narrative against technology and instead develop *new but different technologies*: digital technologies that enable different techno-temporal performances and that are not necessarily merely local. What this means is not so clear. But perhaps we can use, and elaborate on, Borgmann's criteria. The focal practices and occasions he describes are moments when we engage with our environment in a way that uses and develops practical skills and moments that are communal. Borgmann seems to think that we *necessarily* lose this in and with our current use of digital technologies. We miss engagement with the material and with others. We miss common sense and a common time. But this claim can be disputed. Perhaps some technoperformances and their related narratives and processes can also lead to the skilled engagement, focal practices, and common sense-making Borgmann is after. Moreover, his criteria do not take into account context and situation; maybe some situations, also situations in which digital technologies play a role, can suddenly become meaningful occasions, good times. The *kairos*, the auspicious moment, could then in principle also take place when we use digital technologies, when we are involved in a digital technoperformance.

Imagine that I am messaging a friend using an app on my phone, which is usually seen as not really having a good time as opposed to meeting someone in person. We want to be somewhere else, not here and now messaging. We just use it to get across some information. This is not focal. But it is possible that suddenly the messaging takes a direction that renders the exchange focal, that makes both of us not wanting to be anywhere else but here and now, in the digital environment and digital moment constituted by the texting. For example (and to take up a typical existentialist theme), unexpectedly we have a good chat about the meaning of life and death. We could also have had this in person, perhaps, but the point is that it suddenly happens here and now in the technologically mediated environment of digitally mediated messaging. Another example: a digital social medium that is usually seen as a threat to the social life suddenly enables the formation of a community to help people who are in trouble because of the pandemic or the war. Unexpectedly, digital technoperformances create a social environment in which it is good to be in and a *good time*, contributing to good co-existence.

Perhaps we can also find a *common* time in and through digital technologies and our situated digital performances at a level that goes beyond the local and communal, perhaps even at the global level. We could think about how to create more opportunities for *kairos* in and with those technologies within a global technological world. There may be no global or digital equivalent of a campfire. But in addition to enjoying such offline focal moments, we could try to create a global digital environment, digital infrastructure, and in particular digital processes and *technoperformances* in which *kairos* might emerge. I admit that it is not clear yet what this means; for example, we will need to further look into the problems of co-existence and the politics of (common) time making (see below). But in principle the development and use of (new) digital technologies does not preclude the emergence of *kairos* and good time(s).

In any case, it seems that this direction is possible at a personal and local level. We can try to perform time differently by using digital technologies. For example, we can create and perform music with digital technologies, creating opportunities for good *kairos* to emerge. This is more optimistic about the ethical potential of digital technologies than the technodeterministic arguments against clock time, which might miss the power and tyranny of earlier technoperformances (e.g., those performed in a medieval religious context) and which take contemporary technology and its temporalities as given. Instead, we can be more constructive about both digital technologies and time. We can invent new technologies and establish new, better technoperformances of time. But in contrast to technodeterminism of the transhumanist technological future, here uncertainty remains; we never know what will happen. The emergence of the *kairos* cannot be forced. We must remain open and mindful to and in the moment, aware that we are dependent on others, on the environment, and on other temporalities.

Our technologically mediated existence always has a tragic, social, and posthumanist aspect. We perform time but we are never in full control of our (techno)performances. We narrate ourselves, for example by using digital technologies, but we are not the sole authors of our stories. We participate in processes as parts and partners, but we are not in the center. We cannot fully make ourselves. Using digital technologies, we might believe that we are things, but we are not. In process language one can say that we become. In a Heideggerian fashion one might say 'we *ek-sist*'.[16] We are not reducible to an essence, to properties. As we perform, we stand out to the world. We also time out. We are embodied and *ex-posed* to a world that continuously changes, and so do we. Making time and making ourselves also always means *being* made. I perform and I am being performed. Nature co-performs and I, as part of nature, also co-perform. Others co-perform. Technology co-performs. This is so since my performances are always part of, and are made up of, natural, social, and technological processes and interact with performances by others—human and non-human. A process approach to technology is fundamentally relational. As long as I exist, there is no outside of world, there is no outside of process, no outside of narrative, no outside of performance. My becoming is always necessarily worldly and is linked to process, narrative, and performance. To retreat into the singularity of your own 'I' (say, in a neo-Cartesian fashion) or to become an essence means to stop existing, to die. Life is dynamic, moving, relational and processual; the opposite is rigidity, immobility, and death.

To conclude this section: In principle, modern technologies can help to find a better relation to time, rather than precluding that. We should stop ignoring nature's temporalities, we must slow down, but once we understand that we are part of nature and that thinking in terms of processes and performances can also overcome *that* kind of dualism (humans versus nature), we can try to make better times *with*, instead of against, digital technologies and media. We can think of different technoperformances of time.

Finding different technologies that would afford better technoperformances of time is not easy. For example, in response to the problem of modern clock time and the problem of climate change, alternative 'climate clocks' have been proposed. There are clocks that count down to

---

[16] Heidegger, "Letter on Humanism."

catastrophe. For example, in New York a clock by artists Gan Golan and Andrew Boyd shows how much time the planet has left before emission rates will cause catastrophic damage. Yet such technologies are still based on linear clock time and related apocalyptic narratives and eschatological thinking. There are also clocks that relate to other kinds of times. Michelle Bastian has proposed clocks that respond to the temporalities of the climate crisis and, more generally, the temporalities of nature. For example, a clock synchronized with the population levels of sea turtles that may get extinct due to temperature changes—a clock that also tells 'the frustratingly low time of human efforts to respond to recognized environmental threats'.[17] Yet such solutions remain within a narrative that we have to hurry up, that we don't have time; to that extent, it does not really offer an alternative form of temporality. The works mentioned also remain within the apocalyptic narrative that often governs climate discussions. In this sense, they do not fundamentally change the narratives, performances, and temporalities that currently shape our existence. It remains difficult to think, invent, and perform *beyond the clock*.

But bringing about change is not just about technology. If there is no change, perhaps this is so since some people benefit from this inertia, and some more than others. Let us further move into the *politics* of time and, more precisely, the politics of technoperformances of time.

Kairopolitics and Contemporalization: The Politics of Time and Co-existence in Times of Climate Change

The problem of creating better times and common times, if possible good times, is a problem of co-existence and is also *political*. The problem of politics is often defined in terms of finding common spaces or how to be together in places. In the tradition of philosophical republicanism in particular, the idea is that you try to find a common space, a public space, to deliberate and decide about matters of common concern. For example,

[17] Bastian, "Fatally Confused," 44.

Hannah Arendt argued for creating public spaces for political action.[18] And in a more postmodern—he would say: non-modern—vein, Latour also searched for common spaces and co-existence in networks. According to Latour and Hermant, co-existence is a matter of space rather than time.[19] The political challenge is, in Andermatt's words, 'to build a common world-space with co-existing networks'.[20] Latour and Hermant suggest that when history ends 'co-existence is starting'.[21] This rightly puts the question of co-existence on the agenda. But in spite of Francis Fukuyama's famous claim,[22] history does not end and did not end. And we still face the challenge of co-existence. Moreover, one could argue that *co-existence and the political are not only about creating common space(s) but also about creating common time(s)* and public times, and that the challenge today in a time of digital technologies and the networks they create is to build common world-time(s)—albeit perhaps different ones than those offered by global capitalism. Making a common world is then about making a common time, next to making common space. This common time is not necessarily about time as succession, which Latour and Hermant argue against, but about creating the temporal conditions for co-existence. It is not the abstract time of history or (meta)physics but the situated time that has always already been social. We need to look at political ecologies of time next to, or in conjunction with, political ecologies of space. The challenge is not only to conceptualize a relational space of immanence but also a relational *time* of immanence—albeit one that includes a multitude of temporalities instead of one dominant, linear, and totalizing time.

This kind of political time is about co-existence understood as a becoming in which we actively participate. It is not given but is made, performed. To elaborate on the political aspects of technoperformances of time, we can again use Foucault or speech act theory, but now say more about time telling. Taking inspiration from Butler, Derrida, and Austin,

[18] Arendt, *The Human Condition*.
[19] Latour and Hermant, *Paris: Invisible City*.
[20] Conley, "Bruno Latour."
[21] Latour and Hermant, *Paris*, 101.
[22] Fukuyama, "The End of History?"

Bastian sees the act of telling the time as performative. Telling the time is not a statement of fact but a performative act: it 'orders the world in particular ways'.[23] Time telling is thus politically relevant. More generally, time is political in the sense that there are also tendencies towards what we may call *tempocolonialism* or time empires. The times of some people, some cities, and some countries are seen as more important than those of others; there are processes of temporal uniformization, colonization, and disciplining. And this is not just about the performativity of language; (other) technologies and media play a key role in this, ranging from colonial time maps to Western-centered calendars. The performativity of time telling via modern clocks is political. The clock is used to discipline people, workers for example. The 'universal' *chronos* is created by a techno-capitalist system. Global hegemony is not only material and bodily but also temporal. Workers everywhere are subjected to the rhythm and speed of capitalism. There are stories about Amazon workers being forced to urinate in bottles to save time.[24] But also leisure time is regulated. The computer gamer's performances, for example, do not necessarily escape the dictates of clock time. And institutions such as schools and hospitals are still under similar temporal regimes as in the times Foucault described.

Framed in terms of the concepts I proposed: technoperformances with these clocks and other technologies are political. Time and life time are social and political, and so are the technologies that, through performances, narratives, and (other) processes shape our time(s). The concept of technoperformance enables us to point to the political and bodily dimensions of time making through technologies. Through technoperformances, global power relations are translated into what Birth calls 'the physical and psychological temporal experience of the body'. We have to perform in a 24 h/7 economy. This creates problems such as 'the denial of physiology that is crucial for the global functioning of capitalism'.[25] Again, the claim is that biological and natural rhythms are ignored, for example in night-shifts and jet lag. And some people suffer more than others. Birth reminds us that the time that currently organizes us is not

---

[23] Bastian, "Fatally Confused," 32.

[24] Vincent, "Amazon."

[25] Birth, *Objects of Time*, 130.

'natural' or 'real' and is in fact rather recent.[26] Digital technologies contribute to the development, maintenance, and proliferation of such ways
of organizing and performing time. For example, digital technologies
including AI and (other) algorithms are creating new work orders for taxi
drivers and food delivery workers, which are put under regimes of surveillance and control[27]; drivers may be fired if they do not perform well.
Here clock time, in combination with digital technologies, delivers new
forms of capitalist management, exploitation, and colonialism.

But while we may rightly worry about, and protest against, this particular way of organizing time, work, and people, we also have a reason to
be optimistic. If this kind of time is made and performed by humans and
if time is political, then we can change it. Time technologies, including
digital technologies, are not *necessarily* bad, also not socially and politically speaking. Clock time is just one option; there are other options. If
we are out of sync with one another and out of sync with nature, we can
do something about that, with technologies and corresponding technoperformances and narratives. We can transform the social-technological
environment in a way that creates opportunities for focal and synchronous performances, and in a way that gives us back a good kind of present: a present that is connected to the past and the future in a way that
leads to openness, transformation, and life. We can also transform work
orders in ways that end or change specific technoperformative forms of
exploitation and lead to good work and good forms of synchronicity—
and first perhaps synchronicities of resistance. Technologies, in the form
of technoperformances and narrative technologies, can be powerful tools
for such a social and political change. In particular, technologies can
enable specific coordination processes, and performances of collective
performances and collective meaning-making can lead to better co-
existence. If we do not like the times and its time-telling and time-mak-
ing technologies, we need technologies and technoperformances that
help us, collectively, to make better times, to enable good times, and to
syn-chronize in a good way. We want and need better technologies for
better co-existence at local and global level. Better clocks, perhaps, if we

---

[26] Birth, *Objects of Time*, 169.
[27] See for example this study of Uber drivers: Rosenblat and Stark, "Uber's Drivers."

take the broad definition of clocks that Bastian mobilizes (technologies not measuring time but affirming 'a shared social relation to something')[28]. Or we can move beyond clocks altogether and develop co-existence technologies and technoperformances that are no longer clocks and no longer ruled by clock time in any sense, but nevertheless synchronize and organize us in a way that is *good for us*.

Yet we need a discussion about who is included in this 'us'. This needs to involve a discussion about the exclusion of some humans and groups from politics and time (consider again the question: *Whose* time matters?), for example about temporal colonialism. But we may also want to consider recognizing the political status of non-humans. This has been done in animal ethics[29] and environmental ethics, for example, and may be also support by so-called whole earth thinking,[30] which focuses on planetary interrelations and the challenges of co-existence on earth (whereby earth is not just a stage or background for humans but actively participates in human history and vice versa). A better co-existence then means: a better co-existence between humans (for example between people of different cultures and different places and time zones), but also between humans and non-humans, at local and global level. It means a better earth and a better time for all beings on earth, not just humans. We are presently too *anthropochronic*: in our thinking about the present, but also about the past and the future. For example, we imagine the future of humanity, without taking into account the future of non-humans. New technoperformances of time that include non-humans could take us beyond the anthropochronic. We need to connect to times, temporalities, and beings that are not human and that are not made by humans. This is not just about relating to the temporalities connected to our pets, which most pet owners already do (it is part of the care we exercise towards them, for example walking a dog on time in order to take care of its biological needs) but also relating to the temporalities of other living beings, ecosystems, and earth. In the twenty-first century and faced with

---

[28] Bastian, "Fatally Confused," 31.
[29] See for example Donaldson and Kymlicka, *Zoopolis*.
[30] Mickey, *Whole Earth Thinking*, 3.

climate change, we can no longer and should no longer ignore the temporalities of non-human species and non-human nature.

Perhaps we can learn from indigenous cultures, which tend to have different conceptions of time and connect time with the nonhuman and the environment. Interestingly in the light of what I said about performance and bodily movement, this has become especially clear through the observation of gestures, not just language. For example, for the Aymara people of the Andes the past lies in front because it is known and can be seen, whereas the future is behind.[31] And for the Yupno people of Papua New Guinea, time is non-linear and flows uphill: the past flows in the direction of the mouth of the river and the future is towards the river's source.[32] Indigenous narratives also show this nonhuman dimension and different understandings of time and world. As Claire Colebrook writes in her article on stratigraphy of the Anthropocene: 'While the Anthropocene posits a single geological time read through the strata of the earth, there are indigenous cultures that inscribe the space of the earth and its various dimensions within a milieu of nonhuman life (rivers that are the outcome, or depict the shape of, past battles between lizard spirits and bird spirits); even ancient Greek myth placed this world and its human relations after the event of a battle of the Titans'.[33]

Such ancient and indigenous approaches could then be connected to existing frameworks, for example relational and process thinking. For example, the Australian indigenous Bawaka ontology has been described as an ontology of 'co-becoming' 'within which everything exists in a state of emergence and relationality'.[34] Seasons change, people change, trees change, animals change—and we all change and co-constitute each other. The authors rightly argue that this relational ontology of emergence in which a living and interconnected world is continuously produced opens up possibilities for other ways of thinking. Indigenous gestures, views, and narratives may not only help us to understand being 'together in place', as the title of a Larsen's and Johnson's book on indigenous

[31] Nuñez and Sweetser, "Future Behind Them."

[32] Ananthaswamy, "Time Flows Uphill."

[33] Colebrook, "A Grandiose Time," 443.

[34] Bawaka Country et al., "Co-Becoming Bawaka," 456. See also other indigenous relational ontologies, for example Maori or Ubuntu.

coexistence puts it,[35] but also *being together in time*. They reveal how co-existing everywhere and always—not just in places or times linked to indigenous cultures—is both a matter of place(s) and a matter of time(s), and show how co-becoming was always already a relational and more-than-human matter.

This does not mean that co-existence as co-becoming is necessarily harmonious. Places and times are linked to power struggles, and concep-tualizing co-becomings might itself be a political act. For example, in an Australian post-colonial context, Bawaka Country et al. argue that attending to co-becomings 'may shift relationships of power away from an (Anglo) human-centred dominance towards a reconceptualization of a co-emergent world based on intimate human-more-than-human rela-tionships of responsibility and care'.[36] While the authors still frame their effort as an understanding of place/space, it could also be useful to con-sider different ways of thinking about *time* and process in a way that includes 'the vibrant agency'[37] of non-humans and moves away from colonial and imperialist projects. In the colonial political imaginaries and performances, indigenous people were not only erased from the land (which was seen as *a terra nullius*)[38] but also *from time* and history. Like their land, the time of indigenous people was seen as empty, non-historical, indeed non-existent: a temporal nothingness to be filled with the settler's history. Similarly, non-humans are often erased from time. Attending to indigenous ways of understanding—including their under-standing of time—is thus already a tempopolitical performance that puts these humans and non-humans back in time, or rather, that opens up possibilities to give them back *their* time, let them perform their time. However, here too a return to the Garden is neither possible nor desir-able. Neither those who inherit indigenous culture nor others can simply copy pre-modern or local indigenous sense making and politics; we have to create our own, new processes, narratives, and performances of time that connect human with non-human temporalities (and geographies) in

---

[35] Larsen and Johnson, *Being Together in Place*.
[36] Bawaka Country et al., "Co-Becoming Bawaka," 470.
[37] Bawaka Country et al., "Co-Becoming Bawaka," 457.
[38] Richie Howitt calls it 'a politics of erasure.' Howitt, "Scales of Coexistence," 55.

a way that makes sense for us here and now. Furthermore, the point is not that we must choose *one* ontology or narrative. Several relational and process approaches can be combined. As Amanda Lynch and Siri Veland argue, various myths—including that of the Anthropocene itself—can offer partial insights and can be combined and tested in order to navigate coexistence and imagine alternate futures in what they call 'the more-than-human Anthropocene'.[39]

But the natural sciences, too, can help us to move beyond an anthropocentric conception of the Anthropocene and, more generally, of time. As Dipesh Chakrabarty has noted, 'a sense of geological time remains strangely withdrawn in contemporary discussions of the Anthropocene in the human sciences and yields place to the more human-centred time of world history'.[40] By contrast, attending to geological time may help us to attend to time(s) beyond human time(s), in this case time before human time, the earth's deep past. As Chakrabarty reminds us, 'humans come very late in that history'.[41] Geology but also astronomy can help us to what we may call 'contemporalize' (in analogy to contextualize) our own time. Moreover, as Bronislaw Szerszynski argues, both natural 'monuments' and human-made monuments that refer to divine time such as the Parthenon frieze can mediate between human and other temporalities and encourage us to imagine the deep time of earth.[42] We may also consider (other) memory technologies that remind us of the past, both recent and deep: the time of history and the time when history was still silent. Deep time is the silence before the screaming of human time. It is the silence of the rocks, the time of non-human inscriptions in and on the earth. With memory technologies such as tools to measure and study organisms, ice cores, and strata in rocks but also writing, computers, AI and data science, we can try to reveal and mediate those times and let them speak. In addition, the humanities and the arts, with their writing technologies and art media, could assist us to experience the pre-human silence of that deep past or—to give it a Heideggerian-Hölderlinian

---

[39] Lynch and Veland, "Coexistence," 140.
[40] Chakrabarty, "Anthropocene Time."
[41] Chakrabarty, "Anthropocene Time," 25.
[42] Szerszynski, "The Anthropocene Monument."

spin—enable us to listen to the echoes of the voices of the departed gods. With archeological tools and writings, we can also try to connect to the relatively recent ancient history of *human* cultures and communities. Bringing in the humanities, one could complement what Jean Baudrillard called the 'secularization of history',[43] which fixes history in a visible and objective form, with myths and the study of myths—although this too is a technologically mediated revealing and (story) telling. In sum, we need to fully realize, acknowledge, and reveal that *there was time and world before us*: before me, before you, before us (our community, our society), and before humanity. Not only indigenous cultures but also sciences, arts, and technology can help us to connect with these past times, can help us to *contemporalize*.

But also with regard to the future, we can and must contemporalize: we need to become more open to imagining times and timescales that go beyond our life time, beyond the time(s) of our particular communities and societies, and beyond the time(s) performed by humanity. We are too much centered on our own time. Considering and constructing other times will help us to better deal with climate change and will be better for the flourishing of humans *and* non-humans. For example, we should become more aware of the times of other species living today and we should think of the next generations and make sure that they will also have a good time and existence. We thus need *post-anthropochronic* technologies and technoperformances that tell time and make time in different ways, and thereby shape our (current and local) time and existence in a way that is good for us and others now and in the future—human or non-human.

However, views differ considerably about how a more-than-human future may look like. In the context of thinking about future technologies such as AI, another version of post-anthropochronic thinking is *longtermism*, which includes the ethical view that we should give priority to influencing the long-term future. This general orientation towards the future is shared with, say, environmentalists, and rightly rejects ethical presentism. However, longtermism concerns a very different caring about

---

[43] Baudrillard, *Simulacra and Simulation*, 48. Baudrillard writes about film but one could also see archeology as a science that tries to fix history in objective form.

the long term and a very different view of the long term than that of environmentalists. Transhumanist philosophers Toby Ord and Nick Bostrom do not mean that we should care about the earth for the next generations but have a very specific apocalyptic vision of the future in mind, according to which humanity is outdated and life on earth might soon come to an end, but posthuman intelligence has the potential to grow exponentially and spread in the cosmos. According to these transhumanists, this development is something we should support and prepare for (for example when developing AI), instead of engaging in self-destruction. Nature is thus further subjugated and humanity is replaced by an incredibly large number of people from a new species that colonize the universe and perhaps live happy lives in computer simulations. In the light of this technocosmic perspective on the future, earthly and ongoing events like climate change, pandemics, nuclear disasters, or world wars are mere trivialities. We should instead worry about the bigger picture, what Ord in his book *The Precipice* calls our 'longterm potential'[44]: the potential of intelligence spreading and flourishing in the cosmos. For ethics, this potential is what matters, this should not be destroyed. Or to put it in terms of time: the first 100 or 1000 years do not matter; we should focus on the long term.

This longterminism has little to do with the post-anthropochronic and planetary co-existence direction I just articulated. On the contrary, it seems a direct threat to it, as it seems to constitute a hyper-anthropochrone, albeit focused on posthumans rather than current humans. It is not interested in the earth or non-human animals—and indeed hardly in the *humans* that exist today and in the next 100 years. It is not interested in the human and performative-embodied meaning making involved in the projects of contemporalization I suggested, projects which try to relate to other times but take place now and are meant to make meaning now, for us humans. It is also not interested in the future of non-human animals or the earth's ecosystems. Instead, its utilitarian calculations discount our time and the time of the next generations in favor of the zillions of people who might exist in the cosmos one day. Doing something about climate change is then a feel-good project that does not get its priorities right and

---

[44] Ord, *The Precipice*, 6.

is not a form of effective altruism since it does not concern the very long-term effects. Bad luck for us that we find ourselves at the wrong end of the timeline. Surveillance by AI will take care of us, irrational and bug-plagued humans, and superintelligence and its prophets will make sure that the long-term goals are reached. Here thinking about time thus has again ethical and political consequences, in this case very dangerous ones—at least for current humans and non-humans.[45] What matters is the existence of future intelligences. Co-existence of humans and with humans and co-existence with non-humans are not an issue, or at least not the issue we should prioritize. Good co-existence *now* is only valuable if it is instrumental to bringing that particular future of humanity (or rather its overcoming) into being. Longtermism is thus a postanthropo-chronic view that, like the postanthropochronic view(s) I suggested, has clear political significance and consequences—in my view: very problematic ones. But it manages to overcome the usual focus on human time here and now. Anthropochronic thinking has had its day; it is time to move on.

Moreover, even when it comes to humans, thinking about time is often too individualist, methodologically speaking and sometimes politically speaking. I have already suggested that we have the problem of synchronization. We need to find common spaces but also (good) *common times*. We need to move and perform more synchronically. We need to make meaning together. Now to argue for a political performance and synchronization in the sense of finding *common time* is a political statement: today, some people might benefit from *not* doing that and from the absence of common time. The privatization of time works well for them. A specific political ordering and socio-economic system seems to discourage us from finding common time and common meaning, and—through technologies such as AI—disperse us into the multiple "nows" of cyberspace or trap us in the past[46] and the future, rendering focal occasions, common times, and synchronization in the sense of making common time and synchronized resistance against the system difficult if not impossible. I already mentioned the relation between capitalism and clock

---

[45] For a critical view of longtermism see Torres, "The Dangerous Ideas."
[46] Coeckelbergh, "Time Machines."

time. But as my examples suggested, the point is not just about clocks but about a particular way of organization and ordering time and *people*. The clock in Charlie Chaplin's *Modern Times* does not only represent clock time but also organizes the workers in a way that benefits the capitalist factory owner (rather than organizing them in other ways). Digital technologies, at least to the extent that they are embedded within that capitalist structure, seem to have similar effects today, not only in the factory but also in the context of other contemporary digital technoperformances. Capitalism encourages synchronization as economic globalization. But it does not necessarily contribute to, or even hinders, the making of common time(s). Digital media and technologies through which we technoperform time then become tools of capitalism and privatization, and thus create a specific ordering of society and in the end a specific global order. Time is not only money; time is also power. Digital technologies play a crucial role in shaping that time and power.

To further conceptualize this critical perspective on what digital technologies do to our time(s), we need to connect to the political economy heritage of Marx and Engels, which continues to influence contemporary thinking about time. Responding to Virilio and Marx, Jason Adams has argued in *Occupy Time*[47] that presentist and immediatist Internet experience provided the basis for the current political economy. In addition, one may also point to the temporality and political economy of what Wajcman and Dodd call 'the slow digital labor, both paid and unpaid'[48] that powers and hides behind the temporalities of fast use. From a Marxian, critical theory point of view, the question is then how to resist this now-time and (other) capitalist temporalities. Co-existence is not enough; solidarity and resistance are needed to bring about change. But what form can and should this change take? In the spirit of Marx and Engels, one might call for revolution. However, Adams points out that calling for revolution appeals to the same kind of now-time, a time that is still about *chronos*. In other words, we remain stuck in the same time order. He proposes that we produce a 'countertemporality'.[49] Although it

---

[47] Adams, *Occupy Time*.
[48] Wajcman and Dodd, *The Sociology of Speed*, 3.
[49] Adams, *Occupy Time*, 98.

is not clear what this means, it seems to include engaging with specific situations rather than simply slowing down. This reminds us of the *kairos* we already discussed. It is not about changing clock time as such but about doing something together in a situation, which generates meaning and good time. Furthermore, it is interesting that Adams says that his proposal is life affirmative instead of the being-toward-death of Heidegger's existentialism: the idea is to 'give birth to new times'[50]—which of course reminds about Hannah Arendt's notion of natality as the source of our 'capacity to begin.'[51] Hence his proposed solution is *kairopolitical*. But it is not a thinking based on anxiety, as in Heidegger. It suggests a thinking that opens up the new. It is not obsessed with death. It is more Arendtian, it is more about birth. Perhaps it also suggests a more courageous Nietzschean kind of *ek-sistence* that is not based on *Angst* but affirms life.

However, within the current capitalist and (post)modern order, it remains hard to think of new beginnings outside of now-time and clock-time. To say that it is time for change or even for revolution (a performative use of language) can easily be absorbed and disarmed by the capitalist and modern time order. The idea of revolution also does not seem to fit to the idea of slowing down, being mindful, being patient, and so on. Such alternative performances and ancient virtues seem to stand in contrast to those of the impatient modern revolutionary who is ready to seize the political now. They are also all personal attitudes and characteristics; it is not clear what those mean for the collective and for collective action. It remains difficult to conceptualize the link between the personal and the political, in the context of thinking about time and technologies and elsewhere.

One possible direction is certainly to think further about the nature of *political performances*. Neither capitalism nor other political systems exist outside the concrete political (techno)performances of people, and (techno)performances that are not meant to be political may nevertheless have wider political significance. Consider the performances of protest, but also performances of resistance that get less attention such as different uses of technology that maintain or help to establish new forms of

---

[50] Adams, *Occupy Time*, 42.
[51] Arendt, *The Human Condition*, 247.

community and co-existence. Consider hacking performances that have political implications, for example those that hinder the creation of a totalizing capitalist now or sabotage fascist social fragmentation by intervening in Big Tech-organized social media. Or consider communities that manage to retreat from the screaming digital global media order and follow different times and rhythms, for example inspired by Zen Buddhism. In both cases, these performances and related practices can be seen as technoperformances of time that are at the same time political and relate to the concrete lives of people. The concept of performance may thus help to forge conceptual connections between what people do and the wider political and socio-economic structures that shape those performances and are at the same time constituted by them. With regard to time, moving to a different kind of time, making countertemporalities, and forging a different type of synchronicity and co-existence will in any case require different (techno)performances. More work is needed to conceptualize the link between personal performances and larger political structures and processes, also at a global level.

Moreover, given the important role digital technologies such as AI play today, social and political change can only work if it is sufficiently *technopolitical*: if going in a good direction at all, revolution and resistance will also have to use, and transform, the technological basis. Or framed in the terms I proposed: we need to get our technoperformances right. If we really want change, then neither time nor technologies should be accepted as given and unchangeable. Time needs to be technoperformed in a different way—perhaps in a way that does not necessarily reject digital technologies. They may play a role in changing the time(s). Not in themselves (if that makes sense at all) but in the form of technoperformances of time, which always involves people.

Recognizing this role of technology in and as political performance(s) and as crucially involving humans gives an important responsibility to those who develop new technologies. Computer scientists, engineers, and others involved in research, development, and innovation need to be aware that they have an ethical and political responsibility: in general, and also a responsibility for *making time*. Whether we can have better times and good times together will partly depend on the technologies and hence the technoperformances that are now being developed and designed. Those who create the technologies of the future also co-direct and co-choreograph the technoperformances of the future, and thereby shape our time and our existence. In addition, those who use and maintain digital technologies also have an important kairopolitical responsibility, since regardless of what goals and lifetimes those who design and sell digital technologies had in mind, good times with technologies also depend on how we use the technologies, how they are maintained, and what alternative functions we may creatively perform with them. What digital technologies and media are and do depends crucially on its users and participants. For example, social media is not just Big Tech; social media is also you and me. Through alternative technoperformances, we can make better times. This requires synchronicity: the synchronicity of resistance and collective action by workers, for example, but also—at different times—for example the synchronicities of participative design and communal maintenance. Political economy theory tends to be too focused on one moment (labor) within the technoperformative and political temporal whole (and the same is true for the focus on the moment of revolution). Both critical theory and philosophy of technology need to consider a much wider timescape.

Moreover, while Adams's Arendtian (and Nietzschean?) gesture is inspirational, a political economy-oriented Marxian analysis, especially one that is focused on *Capital*, is not sufficiently sensitive to the existential, technological, and environmental dimension of *kairopolitics*: at stake in the politics of time making is not only our social and economic relations and not only humans but also the relations to ourselves, to technology, and to nature. A political economy approach risks to be too anthropochronic and too far removed from the existential and technoperformative challenges we face at personal and collective level in the current global context. The *chronos* of socio-economic history needs to be related to, on the one hand, the *chronos* of climate change and other relevant global and long-term considerations suggested by climate science and other sciences and mediated by digital media (for example about the future of technologies, existential risks related to new pandemics and bioweapons) and, on the other hand, the *kairos* of our good moments and the existential challenges we face as persons and relational beings that participate(d) in these larger processes, narratives, and performances as co-makers of meaning, as technoperformers, and indeed as co-*politicians* who may want to contribute to the emergence of the new. We may be workers and members of a particular society, but we are also co-existers and co-performers of time in many other ways, including ways that (also) connect us with climate-performative processes and other global human/ non-human contemporalities, such as pandemics and extinction of species, which shape our time and experience as *ek-sisting* vulnerable, relational, and political beings.

To conclude, we need new technologies and new technoperformances that not only aim at altering our personal stories and selves but, based on the awareness that what we do with technologies is also always political, also transform the narratives, temporalities, and other structures of the wider socio-political and earthly environment. Creating good times and creating good *common* times is not only about specific moments, individuals, specific virtues, or technologies understood as things. It also concerns changing socially and communally embedded narratives and performances with technology, and the wider socio-political structures and processes in which they are embedded, including temporal-political structures. If we want better times for ourselves and for others, we need a new politics of co-existence and new forms of *synchronicity*, new forms of coordination that are not dominated by capitalism and that are responsive to the temporalities and rhythms outside the modern time order, outside the clock order, and outside the anthropochronic order. Individual change is insufficient and there is no technology-free Garden of Eden available. If we want to change our time, we better try to performatively change those time-orders together and do this not against technology but with technology—including perhaps digital technologies.

What McArthur Mingon and John Sutton say about synchronicity in the context of their analysis of haka, a Māori ritual form of song and dance performance, might be helpful to the further conceptualization of this project of creating new temporalities of co-existence, at least at a local level. Their analysis suggests that skilled performance involving moving together in time and *keeping together* in time facilitates social cohesion, leads to pro-sociality, and stimulates positive affect.[52] As Edward Hall,

---

[52] Mingon and Sutton, "Why Robots Can't Haka."

who did anthropological work on Hopi and Navajo reservations, already showed in *The Dance of Life*, rhythm and time organizes us and ties us together.[53] All this synchronicity is badly needed today for good co-existence. Today we need new dances, new choreographies,[54] and new technoperformances of time.

Perhaps this can be scaled to some extent to larger and more global processes. We need to synchronically and performatively establish new temporal boundaries,[55] structures (e.g., narratives and institutions), dances, and rituals that enable us to escape and preferably replace and re-*time* the 24/7 temporality of capitalism. We cannot go back to the past times, even if we wanted to. But we can perform in a new way, we can create the new out of the past and the present. We can performatively move things and respond to the present challenges: we need better choreographies of time in the light of climate change, the pandemic, war, and other global problems.

As always in politics, for this we have to seize the *kairos*, which here means: the opportune moment. But this is not just a matter of *seeing* when it is time. We need visions, appearances, and revealings, as Heidegger would propose based on his reading of the Greeks (and Kierkegaard and Luther)[56], but seeing always presupposes distance; we also and especially need performances, in particular synchronous, choreographed techno-performances. Heidegger's example is always the craftsperson. But the craftsperson does not only see; she also performs and narrates. She performs *in* time as she makes, say, a piece of pottery. But she also *makes time*, shapes a particular temporality, and relates to others. Our techno-performances need to shape a temporality that is good for us and one that also helps to establish an 'us': not in the sense of an essence but in the sense of a synchronicity. For thinking about kairopolitics in this way, the

---

[53] Hall, *The Dance of Life*. See also Zerubavel, *Hidden Rhythms*.

[54] Choreography is a temporal ordering of performance. The meaning is rooted in the Greek khoreia, which means 'dancing in unison' and from khoros (chorus). In modern individualism, we focus on the protagonists, but forget about the chorus.

[55] Influenced by Erving Goffman's work on roles, Zerubavel has argued that dividing time is helpful for structuring the social life and for social cohesion.

[56] See Critchley, "Heidegger's Being and Time."

metaphor of dance is better: we need to move together, we need to dance time.

This dance should also be more inclusive. We should invite others to the dance or join them: people from other cultures, for example, and non-humans. We could try to get in sync with human and non-human others, finding a common rhythm and building a co-existence by means of new movements. It is not clear yet what this means. We need to think harder about what (good) synchronicity means at the global level and in relation to non-humans. For example, we could start with thinking about how to sync with humans that are far away from us and what we can learn from them about time and temporality. Do others have a better relation to time? Do they better manage to have a good time? Historical research and intercultural, anthropological work, such as that of Mingon and Sutton, is certainly helpful for this purpose. But we cannot and should not romantically go back in time or appropriate other forms of temporality in neocolonial ways. And we can also not just return or be confined to local times. We need to explore ways to achieve forms of local and global non-capitalist, non-nationalist, and non-totalitarian synchronicity that can effectively but at the same time *democratically* deal with the challenge of climate change and other global challenges, including challenges that have to do with coordinating different times, different generations, and different cultures.

Bringing this new time politics into being (or helping it be born) requires political changes, including perhaps not only slow but also fast socio-economic changes; the idea of revolution or fast change should not be rejected *a priori*. Not because we should fear to be too late—this fear stays within the capitalist regime of chronicity—but because we should want to establish the right kind of time at the right moment, yet outside the time order that is given to us, and not so much as time dictators but time finders, time co-performers, and time dancers.

In any case we also need new *technologies*, understood as new processes, narratives, and performances. This is not just a matter of design and development. We users of technology are participants, contributors, facilitators of technoperformances of time. Here evolution and other biological metaphors come to mind.[57] Also metaphors for new beginnings. With regard to making better times, we have to be the midwives of processes of growth and the co-choreographers of new, emerging techno-performances. We have to get better at *timing*, at reading the time, and at performing times, at performing good times. We have to get better at syncing processes, people, and bodies. We have to be better performers-at-risk. We have to write better narratives, but we also have to get better at breathing with technology and at dancing with each other. We need to assist the becoming of new dances, new dances with technology and with each other. The challenge is what times to perform, with digital technologies and other technologies, when we find ourselves in challenging times. Creating and performing new and different techno-choreographies, we may contribute to creating the conditions for new beginnings. We may then hope to be able to face, perform, *transform*, and welcome the new times: the times of climate action, the time after Corona, the time of peace. The challenge is about finding and technoperforming the good times and the common times in times that are changing.

References

Adams, Jason M. *Occupy Time*. New York: Palgrave Pivot, 2014.

Ananthaswamy, Anil. "Time Flows Uphill for the Yupno." *New Scientist* 214, no. 2867 (2012): 14.

Arendt, Hannah. *The Human Condition*. Chicago: University of Chicago Press, 1958.

Bastian, Michelle. "Fatally Confused: Telling the Time in the Midst of Ecological Crises." *Environmental Philosophy* 9, no. 1 (2012): 23-48.

Baudrillard, Jean. *Simulacra and Simulation*. Translated by Sheila Glaser. Ann Arbor: The University of Michigan Press, 1994.

Bawaka Country et al. "Co-Becoming Bawaka: Towards a Relational Understanding of Place/Space." *Progress in Human Geography* 40, no. 4 (2016): 455-475.

Birth, Kevin. *Objects of Time*. New York: Palgrave Macmillan, 2012.

---

[57] See also my book, Coeckelbergh, *Growing Moral Relations*.

Borgmann, Albert. "Cyberspace, Cosmology, and the Meaning of Life." *Ubiquity* Vol. 2007. https://ubiquity.acm.org/article.cfm?id=1232403

Borgmann, Albert. *Technology and the Character of Contemporary Life.* Chicago: University of Chicago Press, 1984.

Chakrabarty, Dipesh. "Anthropocene Time." *History and Theory* 57, no. 1 (2018): 5-32.

Coeckelbergh, Mark. "Time Machines: Artificial Intelligence, Process, and Narrative." *Philosophy and Technology* (2021). https://link.springer.com/article/10.1007/s13347-021-00479-y

Coeckelbergh, Mark. *Growing Moral Relations.* Basingstoke: Palgrave Macmillan, 2012.

Colebrook, Claire. "A Grandiose Time of Coexistence: Stratigraphy of the Anthropocene." *Deleuze Studies* 10, no. 4 (2016): 440-454.

Conley, Verena Andermatt. "Bruno Latour: Common Spaces." In *Spatial Ecologies.* Liverpool: Liverpool University Press, 2012.

Critchley, Simon. "Heidegger's Being and Time, part 8: Temporality." *The Guardian*, July 27, 2009. Accessed November 26, 2021. https://www.theguardian.com/commentisfree/belief/2009/jul/27/heidegger-being-time-philosophy

Donaldson, Sue, and Will Kymlicka. *Zoopolis: A Political Theory of Animal Rights.* Oxford: Oxford University Press, 2011.

Foucault, Michel. *Discipline and Punish.* New York: Vintage Books, 1977.

Fukuyama, Francis. "The End of History?" *The National Interest* 16 (1989): 3-18.

Gould, Stephen Jay. *Time's Arrrow, Time's Cycle.* Cambridge, MA: Harvard University Press, 1987.

Hall, Edward. *The Dance of Life.* Garden City, NY: Anchor Press/Doubleday, 1983.

Heidegger, Martin. "Letter on Humanism." In *Basic Writings: Nine Key Essays, plus the Introduction to Being and Time*, edited by David Farrell Krell. London: Routledge, 1978.

Howitt, Richie. "Scales of Coexistence." *Macquarie Law Journal* 6 (2006): 49-64.

Jung, Sang In, Na Kyung Lee, Kyung Woo Kang, Kyoung Kim, and Do Youn Lee. "The Effect of Smartphone Usage Time on Posture and Respiratory Function." *Journal of Physical Therapy Science* 28, no. 1 (2016): 186–189. https://doi.org/10.1589/jpts.28.186

Kreps, David. *Against Nature.* New York: Routledge, 2018.

Larsen, Soren C., and Jay T. Johnson. *Being Together in Place: Indigenous Coexistence in a More Than Human World.* Minneapolis: University of Minnesota Press, 2017.

Latour, Bruno, and Emilie Hermant. *Paris: Invisible City*. Translated by Liz Carey-Libbrecht. 2006. Available at http://www.bruno-latour.fr/virtual/PARIS-INVISIBLE-GB.pdf

Lynch, Amanda, and Siri Veland. "Coexistence." In *Urgency in the Anthropocene*, 139-162. Cambridge, MA: MIT Press, 2018.

Mickey, Sam. *Whole Earth Thinking and Planetary Coexistence*. New York: Routledge, 2016.

Mingon, McArthur, and John Sutton, "Why Robots Can't Haka: Skilled Performance and Embodied Knoweldge in the Māori Haka." *Synthese* 199 (2021): 4337-4365.

Nuñez, Rafael E., and Eve Sweetser. "With the Future Behind Them: Convergent Evidence From Aymara Language and Gesture in the Crosslinguistic Comparison of Spatial Construals of Time." *Cognitive Science* 30, no. 3 (2010): 401-450.

Ord, Toby. *The Precipice*. New York: Hachette Books, 2020.

Proust, Marcel. *In Search of Lost Time*. New Haven: Yale University Press, 2013.

Rosenblat, Alex, and Luke Stark. "Uber's Drivers: Information Asymmetries and Control in Dynamic Work." *SSRN Electronic Journal* 10, no. 27 (2016). https://doi.org/10.2139/ssrn.2686227

Szerszynski, Bronislaw. "The Anthropocene monument: On relating geological and human time." *European Journal of Social Theory* 20, no. 1 (2017): 111-131.

Torres, Phil. "The Dangerous Ideas of 'Longtermism' and 'Existential Risk'." *Current Affairs*, July 28, 2021. Accessed 1 May 2022. https://www.currentaffairs.org/2021/07/the-dangerous-ideas-of-longtermism-and-existential-risk

Vallor, Shannon. *Technology and the Virtues*. New York: Oxford University Press, 2016.

Vincent, James. "Amazon denies stories of workers peeing in bottles, receives a flood of evidence in return." *The Verge*, March 25, 2021. https://www.theverge.com/2021/3/25/22350337/amazon-peeing-in-bottles-workers-exploitation-twitter-response-evidence

Wajcman, Judy, and Nigel Dodd, eds. *The Sociology of Speed*. Oxford: Oxford University Press, 2016.

Zadeh, Joe. "The Tyranny of Time." *Noema Magazine*, June 3, 2021. https://www.noemamag.com/the-tyranny-of-time/?fbclid=IwAR2X5jL8KvYSfBwS8-3o6xVh6PzG8RYxfQ3AnMabWPN-xWrT6jxKZl8gjOU

Zerubavel, Eviatar. *Hidden Rhythms: Schedules and Calendars in Social Life*. Chicago: University of Chicago Press, 1981.

Index[1]

---

[1] Note: Page numbers followed by 'n' refer to notes.

© The Author(s), under exclusive license to Springer Nature Switzerland AG 2022
M. Coeckelbergh, *Digital Technologies, Temporality, and the Politics of Co-Existence*,
https://doi.org/10.1007/978-3-031-17982-2